[signature]
1997

A Retailer's Guide to Glass and Pottery

A Retailer's Guide to Glass and Pottery

Kenneth Blakemore
Editor of *Watchmaker, Jeweller and Silversmith*

Butterworths
London Boston Durban Singapore Sydney Toronto Wellington

All rights reserved. No part of this publication may be reproduced
or transmitted in any form or by any means, including
photocopying and recording, without the written permission of
the copyright holder, application for which should be addressed to
the Publishers. Such written permission must also be obtained
before any part of this publication is stored in a retrieval system of
any nature.

This book is sold subject to the Standard Conditions of Sale of
Net Books and may not be re-sold in the UK below the net price
given by the Publishers in their current price list.

First published 1984
© K. Blakemore, 1984

British Library Cataloguing in Publication Data

Blakemore, Kenneth, *1918–*
  A retailer's guide to glass and pottery.
  1. Glassware   2. Pottery
  I. Title
  748.2   NK5106

ISBN 0-408-01219-6

Library of Congress Cataloging in Publication Data

Blakemore, Kenneth.
  A retailer's guide to glass and pottery.

  Bibliography: p.
  Includes index
  1. Glassware.   2. Pottery.   I. Title
  TP865.B53 1983   666   83-21047
  ISBN 0-408-01219-6

Photoset by Butterworths Litho Preparation Department
Printed in England by MacKays of Chatham

Preface

This book has been written to provide those whose job it is to sell domestic glass and pottery with information about these products which, hopefully, will give them a greater appreciation and a better understanding of what they are dealing with. More than one manufacturer has suggested to me, during my research for this book, that if everyone engaged in the selling of fine glassware and pottery could make frequent visits to the factories where these are produced, it would greatly benefit the entire trade. Those who saw the skills that are involved in producing a cut crystal wine glass or a bone china plate and the care that is taken in the leading factories to maintain standards would, it was felt, pass on the knowledge and the enthusiasm they had acquired to the public. They would also be better equipped to justify price, answer queries and deal with complaints. Unfortunately this is not likely to happen. Only a minority of those engaged in selling these products will participate in even a brief tour of a factory. Many sales-people may therefore find themselves less well informed than some of their customers, for nowadays the public visits factories such as Wedgwood and Stuart in tens of thousands. It is hoped, therefore, that this book will provide an alternative source of information for those who do not have the opportunity to see the products being made for themselves. I believe that it may also assist those who do have this opportunity to gain a better understanding of what they are witnessing. I have found that even though I have been able to tour factories accompanied by technical experts, it has taken me many visits to understand everything that was going on and to grasp the degree of skill which the craftsmen involved in these industries bring to their work.

Glass and pottery have a good deal in common; both are produced by fusing silica, and both presented the pioneers of the crafts with similar problems. The early glass-makers and potters had to discover how to temper the silica so that they could melt it at temperatures that could be reached in a wood-burning furnace, and each craft has had to evolve a complex chemistry over the centuries in order to make the refined wares we now take for granted. Glass and pottery serve similar purposes. They provide us with tablewares that are both practical and attractive and decorative wares for our homes. Both have a host of other applications which are outside the scope of this book. But it would be difficult to conceive of a world without the benefits these materials provide, a world without window glass or sanitary wares, to take just two examples.

Many books have been written about antique glasswares and pottery and I have made no attempt to provide a guide to the products of the past. It is, however, essential for a complete understanding of these two crafts to have some knowledge of how the technology has developed. The industries which today produce fine stemwares and dinner services are still very much craft industries, and the skills involved have been handed down from generation to generation, largely unchanged, though considerably refined with the passing of centuries. Also many of today's designs are rooted in the past, and not a few novelties turn out to be based on the innovations of glass-workers who practised their craft in Murano or potters who worked in China many centuries ago. It is for these reasons as well as to give those who read this book a sense of history that I have traced the two crafts from their beginnings thousands of years ago to the present day.

The remainder of the book is given over to descriptions of how glassware and pottery is produced and decorated in modern factories. In the course of these chapters I have emphasized just how labour-intensive these industries are, and how much skill is involved in producing such everyday things. I have been at pains to stress the great care which the leading producers take to ensure that what they send out of the factories is 'perfect'. Anybody who visits these factories cannot but be impressed by the exacting standards of inspection. Nowadays fine pottery and fine glassware is anything but cheap, but if one knows just what is involved in making these wares one is surprised they are not more expensive.

I have also sought to explain why there are sometimes disparities of price, the reason for which is not always obvious. Why is it likely, for instance, that a cut glass stemware with very few facets on it will be more expensive than one that is elaborately cut, or why is one pattern of porcelain hand-painted under the glaze twice as expensive as another apparently similar pattern? I have also discussed the reasons that wares are sold off as seconds, because only by understanding the common faults can those who work in the retail trade come to appreciate a perfect product when they see one. There are other things that are not always understood. The public are certainly not always aware that six hand-made and hand-cut wine glasses will never be absolutely identical. Machines may be able to produce identical products, but if you watch a foot-blower at work and then watch the gaffer create stems and feet from a gather of glass from the furnace, then follow these glasses to the cutting shop and watch the cutters working freehand within hatched guidelines, you can well understand why a manufacturer of hand-made glasswares has to accept small tolerance on size and weight and minor deviations in the pattern. Hopefully, I have also provided sales-people with the answers to such questions as: What is the difference between English and Continental lead crystal and what is semi-lead glass? What is in-glaze decoration on china and why is the colour range limited on under-glaze decorated wares?

The final chapter of this book consists of a series of brief profiles of those firms whose wares are likely to be found in a shop selling a cross-section of the products of the two industries. This is not intended to be a complete buyer's guide to the industry, as there must be many hundreds of firms, large and small, involved in it today. A number of the firms included have been most cooperative in providing me with information. Indeed, I could not have written this book without the help of those companies and the department heads and craftsmen who have given so freely their knowledge and their time. I owe a particular debt to the staff of Stuart Crystal, Stevens and Williams, Royal Copenhagen, Rosenthal, Royal Doulton, Royal Worcester and Wedgwood.

KENNETH BLAKEMORE

Contents

Preface

1 **Glass-making down the ages** — 1
The first vessels – The birth of blowing – The glasswares of Rome – The importance of Venice – Glass-making in England – The eighteenth century – The nineteenth century – The revival of cutting

2 **Making glasswares** — 25
Raw materials – Making the pots – Filling the pot and firing – Inside a glasshouse – Making a wine-glass bowl – Machine-made glasswares

3 **Decorating glass** — 43
Glass on glass – Balusters and twists – Cutting and engraving – Sand-blasting – Price justification

4 **The discovery and development of the potter's craft** — 60
The beginnings of science – The potter as artist – The discovery of porcelain – Hard-paste porcelain – Soft-paste porcelain – Wedgwood and jasper ware – The nineteenth century – The twentieth century

5 **Making pottery today** — 86
Clays, bodies and glazes – Preparing the ingredients – Making plates – Making hollowares – Throwing and coiling

6 **Decorating pottery** — 105
Decorating and price – Hand-painted pottery – Under-glaze painting – In-glaze decoration – Transfer printing – Lithographic printing – Gold-enrichments – Acid gold – Gold-paste decoration – The role of the inspector

7	Pottery figurines	126
	A century of contrasts – Moulding techniques – The designer – Materials and methods – Painting	
8	Advice on caring for pottery and glass	132
	Looking after bone china and porcelain – The care of glasswares	
9	The industry today	137
	Baccarat – Bohemian glass – Royal Copenhagen – The Dema Glass Group – Royal Doulton – Royal Grafton – Holmegaard – Lalique – Nuutajarvi Glass – Orrefors – Poole – Rosenthal and Thomas – Rye Pottery – Stevens and Williams (Royal Brierley) – Stuart Crystal – Waterford – The Wedgwood Group – Royal Worcester	

Select Bibliography	171
Glossaries	173
Glass	173
Pottery	178
Index	187

1 Glass-making down the ages

It would be difficult to imagine a world without glass. There would be no panes in our windows, no mirrors on our walls, no electric light bulbs. Photography would not have been invented, nor television. Science would have progressed more slowly if the chemist had had no test tubes and the physicist no microscope. And just imagine a dinner table devoid of the glint of those glasswares that allow us to appreciate the subtle colours of wines and enjoy to the full their fragrant taste.

All these, and the many other boons that glass has brought us, we owe to some man in the distant past who stumbled, perhaps quite by accident, on a method of fusing silica sand. Nobody knows for certain when this happened, but we believe that the first man-made glass was produced either in Syria or in Egypt, probably as early as 3000 BC. Some authorities, however, favour a date as much as a thousand years earlier than this. The evidence that exists is scant to say the least, consisting of no more than a few glass beads and some fragments of tiles with glazed surfaces.

The first vessels

The first known glass vessels to come down to us are two little jugs and a beaker bearing the cartouche of Tuthmosis III, which we can therefore date with certainty to the middle of the second millennium BC. The footed cup, in the form of a lotus flower, is fairly sophisticated, and suggests that the technique of making glass and fashioning it was well developed by this time in Egypt. By this period, too, it seems, that glass-making was a well-established craft in Mesopotamia, that other cradle of civilization. A glass bottle has been found at Ur, the opaque body of which has been decorated with blue trails, indicating a high degree of technical knowledge on the part of the man who made it. Further evidence of the existence of the craft of glass-making on a considerable scale at this time is provided by the discovery made by Sir Flinders Petrie, at the end of the nineteenth century, of the the earliest known glasshouse. He found this at Tell el Amarna in Egypt, and the date of it has again been established as the middle of the second millennium BC.

Solving the problems

How man came to discover glass-making is one of those minor miracles of the past. It was not just a matter of fusing sand. The melting temperature of silica sand is very high, and even today the glass-maker has to mix his sand with other materials to reduce the temperature at which it fuses. A further additive was also needed to stabilize the glass. How, one wonders, did man in ancient times stumble on the solutions to these problems?

The Roman historian Pliny, writing in the first century AD, attributes the invention of glass to chance. He recounts how some Phoenician sailors beached their ship in the estuary of the river Belus that lies at the eastern end of the Mediterranean. The cargo of the ship happened to consist of blocks of natron, a form of sodium carbonate, and when the sailors wanted to cook themselves a meal they gathered wood, lit a fire and used some of the natron blocks to support their pots. They then fell asleep, and in the morning they found around the edges of the fire a strange hard translucent substance glittering in the morning sun. Chance had brought together the essential elements of glass-making – fire, silica sand and the natron from the ship, the very material that the earliest glass-makers undoubtedly used to temper the sand, so that the heat given off by burning wood was sufficient to melt it. Roman history tends to be a *mélange* of fact and fable; this story is a way of explaining the inexplicable, and for that reason would have appealed to Pliny. Unfortunately a mixture of silica sand and natron alone would have produced an unstable material similar to isinglass. The missing constituent necessary to produce common glass was lime, and it is difficult to see how this could have been present on the site of that campfire on that beach long ago. It seems not unlikely, however, that an accident pointed the way, and that man eventually succeeded in making stable glass as a result of endless trial and error.

Natural glass

Strangely enough, man was making use of glass many thousands of years before he succeeded in making it. Early man-made tools and weapons from the natural glass obsidian, which was formed as the result of volcanic action fusing silica millions of years ago. But it seems doubtful whether man-made glass would have resulted from a conscious effort to simulate this natural substance. Another natural silicate, the colourless form of quartz which we call rock

Fig. 1.1. An Egyptian glass vase and jug from about 1475–1375 BC (by courtesy of the British Museum)

crystal, seems to have provided a goal for glass-makers at a later period, leading them to search for a clear and colourless material similar to that which nature had created. But this was still a long way in the future. The early glasswares which can be seen in museums today are a far cry from pure, limpid rock crystal (Fig. 1.1).

The problems

There is no doubt that the early glass-makers had a desire to improve their product. This is something inherent in the make-up of a craftsman, but the glass-maker faced two problems which were not to be solved for thousands of years. One was that even a very small amount of impurity in the raw material resulted in his glass being both opaque and off-colour. The other was the problem of fashioning the treacly 'metal' from his furnace into a hollow vessel. With the techniques at his command he could do this only with difficulty, and he was restricted as to size. Few early glasswares are more than four inches in height.

To begin with, the glass-makers were probably not very selective in their choice of sand. Chemical analysis has revealed as many as eighteen different constituents present in their glass, most of which can probably be attributed to the fact that the raw materials they used, the sand in particular, was very impure. At that time nobody, in all probability, appreciated the need for purifying these ingredients, and even if they had it is doubtful whether the means existed to do this effectively. It needs only 0.03 per cent of iron to be present to give glass a green or yellow tinge. How important it is to start with the right sand is indicated by the trouble the modern glasshouse goes to obtain this. Today's glass-makers prefer inland sand to that from the seashore, because the sand from places like Loch Aline in Scotland or Fontainebleau in France is more even in grain and contains fewer impurities. The glass-makers clean this sand scrupulously and then add decolourizers to their mix to take care of any metallic oxides that have remained in it. The soda or potash which the ancient glass-makers used was also probably very impure, and so, too, would have been the lime they recovered from the limestone rocks. It is hardly surprising, therefore, that the metal the ancient

Egyptian glass-workers produced bore so little resemblance to the metal in the pots in a modern glasshouse.

Before blowing

The technique of blowing glass, which made it possible to produce glasswares to rival the products of the potter's craft, was not invented until 50 BC. Until that time the most effective way to make glass vessels was what we call the core technique. What the ancient glass-maker did was to build up wet sand, or more probably clay, on the end of a metal rod. He would then fashion this into the required shape, dip it into the molten glass in his furnace and roll the glass on a stone slab to even it out, the process of 'marvering', still practised by the glass-maker today. When the glass had cooled and hardened he would remove the rod and shake out the sand, or clay, core. He had now produced a hollow form. If he needed to provide his vessel with a foot he could have added a second gather from the furnace before removing the rod and shaped this additional 'metal' with simple tools, as the feet of glasswares are still formed to this day. Another method that we believe he used was to cast a vessel in two parts by pouring glass into hollow moulds and then fusing the two halves together, a technique still used for making cheap pressed glasswares. Virtually the only form of decoration which we find on early glass is what we call 'trailed decoration'. To produce this, the glass-maker would have in his furnace a second pot of 'metal', deliberately coloured by the addition of some metallic oxide. He would dip a stick into this and trail a thread of this coloured glass over his vessel before this had cooled. This was how glasswares were made for thousands of years.

The birth of blowing

By the end of the first millennium BC the glass-making industry seems to have become centred in Phoenicia, and it was almost certainly in the city of Sidon in 50 BC that some glass-worker discovered that if he gathered his glass on the end of a hollow pipe and then blew through the pipe he could blow out the glass, as a child blows a soap bubble, to produce a delicate sphere. The glass-makers could now produce larger vessels. Examples 20–25 cm (8–10 inches) high began to be produced in the first century BC, and they were far more attractive than the vessels produced by the core technique, because blown glass acquires a natural and beautiful curvature. At this time, too, the glass-makers began to blow their glass in moulds, often patterned. So the techniques of glass-making which are still employed today were born. All that has been achieved in the centuries since then is that we have learnt to produce better 'metal' and have refined the ancient techniques of making vessels from it.

The glasswares of Rome

Inevitably, the Romans took up glass-making, almost certainly importing craftsmen from Phoenicia, who then passed on their skills to the Roman craftsmen. The Roman industry was at first centred in Campania, where suitable sand was available at the mouth of the river Volturno, but as early as 14 AD a glasshouse had been established in Rome itself. The Romans' greatest single achievement was to apply to glass-making the technique of cameo-cutting. Cameos cut from agates from Sicily and from Idar Oberstein in Germany were in great demand in Rome, and glasswares decorated in a similar way were calculated to appeal to citizens enriched by Roman expansion throughout Europe and the Middle East. Only a single example of this cameo-cut glass survives, the famous Portland Vase (Fig. 1.2), but this accomplished piece of craftsmanship can hardly have been the only one of its kind. The Portland Vase was brought to London in 1785 by the British Ambassador to the Court of Naples, Sir William Hamilton. It was subsequently acquired by the Duke of Portland, who in 1810 lent it to the British Museum, who eventually purchased it in 1945. In the meantime, however, in 1845, it had been smashed into about two hundred pieces by a certain William Lloyd, and had had to be painstakingly reconstructed. The vase was blown from blue glass, made by adding cobalt. Over this, opaque white glass had been overlaid. This white glass had then been carved to reproduce in exquisite detail the various figures present at the mythical marriage of Peleus and Thetis and the foliate motifs that decorate the vase.

Fig. 1.2. The Portland Vase, in blown and carved glass, from the first century AD (by courtesy of the British Museum)

8   Glass-making down the ages

Forest glass

Besides establishing new standards in glass-making, exemplified by the Portland Vase, the Romans were also responsible for the spread of the glass-makers' craft throughout Europe. Glass-works were established in many parts of the Roman Empire, notably in

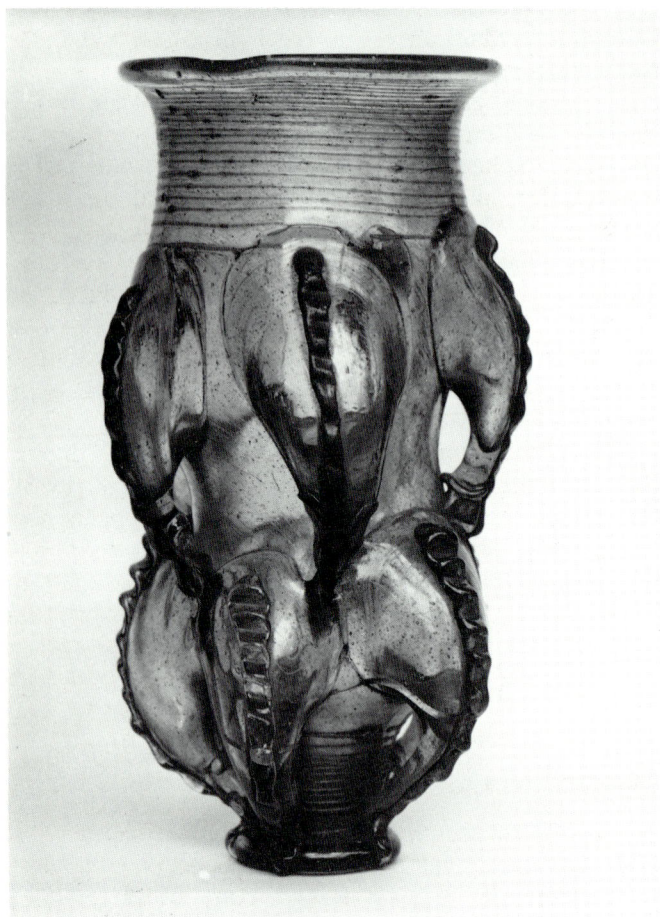

Fig. 1.3. Anglo-Saxon glass claw beaker from Castle Eden, Durham (fifth or sixth century AD) (by courtesy of the British Museum)

the Seine–Rhine region, where a metal known as waldglas (*wald* meaning forest) was produced over a long period. The potash used by the makers in this area was unrefined ash resulting from the burning of vegetable matter, and the impurities in this ash produced a green, yellow or brown metal which remained fashionable in Germany long after the technique of making colourless glass was well understood there. Indeed, the wine glasses favoured in the Moselle region are still green in colour, and their style, with their heavy knopped stems, harks back to the waldglas tradition. Many of the glasshouses established under Roman rule did not survive the fall of the Roman Empire. In Britain, for instance, glass-making ceased after the Romans left these islands. The German industry, though, certainly seems to have survived the upheavals of the Dark Ages, and continued to produce glasswares, in a style already established by the fourth century AD, right through the Middle Ages. Besides their characteristic colour these glasswares were decorated in a distinctive style which exhibited a considerable command of the glass-blower's art. The feature of these wares were the prunts used to decorate bowls and stems. These were blobs of glass applied to the body of the vessel, sometimes decorated, like the raspberry prunts the surfaces of which are covered with seed-like blobs. Sometimes these blobs were blown from inside the glass and drawn out so that they resembled claws or dolphins (see Fig. 1.3). During the Middle Ages this form of decoration was applied to a whole range of utilitarian glasswares, indicating that by this period in Germany glasswares were widely used on the table for drinking wine. These drinking glasses included the 'humpen', a generous beaker, the tall and slender 'stangenglas' and the 'romer', which with its wide stem is the ancestor of the modern German wine-glass.

The importance of Venice

Meanwhile in the thirteenth century glass-making had become established in Venice, and the Venetian glassmakers were to revolutionize the craft. They set new standards which created a growing demand for glasswares all over Europe which, in turn, led to a great expansion of the industry. However, the immediate popularity of Venetian glass created problems. By 1291 the number of glasshouses in the city had created what the authorities considered to be an unacceptable fire hazard, and an edict was passed ordering the glass-makers to remove their furnaces to the offshore island of Murano. To reduce the possibility of competition developing which might adversely affect the industry's thriving export business, the workmen employed in the Murano glasshouses were forbidden to leave the island on pain of death. Needless to say, there were those who were prepared to take the risk, and by the seventeenth century the secrets of glass *à la façon de Venise* had been disseminated throughout Europe.

Venice's two main contributions to the development of glass-making were to set new standards for the metal and to establish a new form of glassware – the stemware which today accounts for the bulk of the output of all the major glasshouses. The new metal for which Venice became famous was known as 'cristallo' because of its resemblance to rock crystal. This was introduced in the fifteenth century, and was made by grinding quartz pebbles from the bed of the river Ticino, which provided a purer source of silica than sand. The Venetians also introduced manganese, a decolourizer which neutralized any residual colouring agents and produced a clear and almost colourless metal. Despite a determined effort to obviate impurities and the use of a decolourizer, the Venetians were not completely successful. Cristallo has a faint greyish tinge. The glass-makers found, however, that by blowing the metal very thin this tinge became almost imperceptible. It might be said that this slight shortcoming of cristallo dictated the delicate style of Venetian glass which was to make it famous.

The Venetians' other great innovation was the introduction of a slender stem between the bowl and the foot of their glasses. Prior to this, with rare exception, drinking vessels had taken either the form of a footed bowl or a footed beaker. Not only did the Venetians introduce the stem, but they elaborated it so that it became the focus of the decoration. They exploited the ductility of glass to create a fantasy of trailed decoration around the stem, which, as in the case of the famous winged glasses, made it quite impossible to hold the glass by its stem as one usually does today. This concentration on applied decoration was, of course, a direct result of the necessity to blow the glass thin to disguise that grey tinge. It would have been quite impossible to engrave or carve such thin material. Apart from trailed decoration, the other main forms of decoration employed by the Venetians were gilding and enamelling.

These ingenious glass-makers living on their island did, however, also develop other techniques for producing decorative wares, techniques that have been revived in more recent times.

Fig. 1.4. A Nailsea jug of green glass, with threads of opaque white (by courtesy of the Victoria and Albert Museum)

The most dramatic of these was latticinio, which consisted of fusing opaque white glass and cristallo together to produce banded wares, the bands sometimes being of a threadlike fineness, so producing a lace effect. The Nailsea style of glassware, produced in a number of English factories at the beginning of the nineteenth century, exploited this same duotone technique (Fig. 1.4). Another Venetian achievement which was revived in the nineteenth century was millefiori decoration. This was produced by fusing together rods of different coloured glasses to produce a cane. This cane was then cut into short lengths which were covered with a gather from the furnace, and when the ware was formed polychromatic effects were produced. This technique is, of course, still used today to produce millefiori paperweights.

The next significant development took place in Bohemia, still one of the world's major centres of glass-making. By substituting potash for soda in the seventeenth century, the Bohemian glass-makers produced a metal appreciably more brilliant than cristallo.

In northern Europe the Latin exuberance of the Venetian style was translated into more workmanlike designs, and this tendency was exemplified by the wine glasses made in Holland in the middle years of the seventeenth century. The Dutch glass-makers borrowed the baluster motif from baroque architecture and produced the baluster stem, which was at once decorative and practical and which the English glass-makers were to adopt and use to such good effect throughout the first fifty years of the eighteenth century.

Glass-making in England

Let us now, however, look back briefly at the early history of glass-making in England. Undoubtedly, glasswares were made in this country during the period of the Roman occupation as they were throughout the Roman Empire. When the Romans left English shores, however, the craft seems to have died out entirely, and it was not to be revived for a long time. We did not, it seems, even produce window glass in England, for the Venerable Bede, writing at the beginning of the eighth century, recounts that

Benedict sent messengers to Gaul to fetch over glass-makers to glaze the windows of his church. Some fifty years later, in 758, the Bishop of Jarrow was writing in desperation to the Bishop of Mayenne requesting that 'if there be any man in your diocese who can make vessels of glass, I beg your fraternity that you will persuade him to come to us, for we are ignorant and helpless in the art'.

The foundations of the industry

It seems, indeed, that it was not until the thirteenth century that any quantity of glass was produced in England. By this time a glass industry had become established at Chiddingford in Surrey, and in 1226 a man called Laurence Vitrearius, who probably came from Normandy, set up a glasshouse there. In 1240 he obtained the commission to provide the glass for the windows of Westminster Abbey. Glass-making seems to have been carried out continuously in this rather unlikely area up to the middle years of the fifteenth century, but the production of the Surrey glasshouses was mainly restricted to window glass and bottles. Then towards the end of the fifteenth century a number of skilled workers came over from Venice and Lorraine to practise their craft in Surrey. Among these emigrant craftsmen was Jean Carré, who had previously run a glasshouse in Antwerp.

Jean Carré subsequently moved his operation to London in 1570, where he established a 'cristallo glasshouse' in Crutched Friars and brought over families of glass-workers, mainly from Lorraine. These families included the Hennezels, the Houxs and the Thietrys, and they were to lay the foundations of the English glass-making industry that exists today. Their names, somewhat anglicized, crop up in various parish records, and the 'Henzels', the 'Tysacks' and the 'Titterys' established the craft of glass-making in Stourbridge, Bristol and Newcastle-on-Tyne.

Carré obtained the first of a series of monopolies granted to foster the craft by encouraging home-produced wares at the expense of imported ones. On his death in 1572 he was succeeded by a Venetian, Giacomo Verzelini, who had been managing the Crutched Friars glasshouse. Verzelini obtained an exclusive license from Queen Elizabeth I to produce cristallo in the

Venetian style in England, but was expected to teach his skills to native craftsmen in exchange for this privilege. During his seventeen years at Crutched Friars he set new standards for English glasswares and is generally considered to be the father of the craft in England. Also because of his habit of signing and dating glasses the products of his factory are the first English glasswares that can be assigned to a specific maker (see Fig. 1.5).

Fig. 1.5. A goblet from the London glasshouse of Giacomo Verzelini, dated 1581 (by courtesy of the Victoria and Albert Museum)

The coal monopoly

The next important event in the history of English glass-making was the application for a patent by Sir Edward Zouch, Boris Thelwell and Thomas Percival, covering the design for a coal-fired furnace suitable for use in a glasshouse. A patent for twenty-one years was duly granted in 1614, and it was also ordained that wood should no longer be used for fuelling the furnaces in glasshouses. In outlawing the use of wood the government of the day were concerned to encourage the use of coal in industry with a view to conserving timber, which was so important for ship-building. This prohibition was, though, to have an unexpected benefit for the glass industry. It has indeed been described as 'one of the great turning-points in the development of the English glass trade'. One of the problems which early glass-makers had to contend with was what is known as 'seeding'. As the constituents of the metal fused in the pot in the furnace, gas bubbles were created. Unless these were brought to the surface they would be present in any vessel made from the metal and so mar its appearance. The higher temperatures achieved by firing the furnace with coal and the greater ability to regulate a coal-fired furnace, though they did not entirely obviate the problem of seeding, certainly reduced it. Yet another spin-off from the adoption of coal-firing was the change it wrought in the structure of the industry. The patent-holder for the furnace now controlled glass-making, and by good fortune the patent came into the hands of a man with both foresight and ambition, a retired naval officer of some distinction, who had been knighted for his services, promoted to vice-admiral and in 1604 had became Treasurer to the Navy. This obviously able man, Sir Robert Mansell, joined Sir Edward Zouch's firm in 1617 and soon afterwards bought out the other partners, and so obtained sole control. For the next quarter of a century Sir Robert Mansell was to reorganize and expand the English glass industry. It made sense to him to establish glasshouses in areas where coal and clay, from which to make pots, was available and he licensed glass-makers to establish glasshouses in the West Midlands, Staffordshire, Shropshire, Nottinghamshire and Newcastle in return for paying him royalties on their production. He also controlled the distribution of their wares.

Lead in the metal

The next key figure in the development of glass-making in England was George Ravenscroft, who had started a glasshouse in London at the Savoy in 1673. Mansell's monopoly, though it had led to a rapid growth of the industry, had resulted, because of the lack of competition, in the products of the new glasshouses being of a low standard, certainly compared with those produced in Venice. To organize a protest against this monopoly, the Company of Glass Sellers was set up in 1635. They do not seem to have become much of a force in the trade until after the restoration of the monarchy in 1660, when they became active in importing glass from the Murano glasshouses. To ensure that this Venetian glass accorded with English tastes, designs by John Greene and other English designers were sent out to Venice. However, problems arose over the interpretation of these designs, and this prompted the Company of Glass Sellers to encourage the production of fine glasswares in England. It was at this point that Ravenscroft became involved with the Glass Sellers. As the result of a long period of experiment, he produced a glass with lead oxide as one of the major ingredients. He therefore made the first lead crystal, for which English glass-makers have been famous for nearly three hundred years.

Not only did the addition of lead oxide to the constituents of glass, replacing the soda content, greatly improve the appearance of the glass, but it went a long way towards overcoming one of the major drawbacks to soda glass at that time – its inclination to crizzle. Crizzling is like a disease of the glass. Internal cracking develops, a sour-smelling liquid forms on the surface and the glass gradually decomposes and eventually crumbles away. Glass-makers had experimented with lead oxide prior to this, but it had not provided a cure for crizzling. Ravenscroft, however, persisted, and eventually, once he had got the proportions right, found the cure. In 1676 the Glass Sellers issued a document to the effect that: 'We do certify and attest that the defect of the flint glass, which was formerly observed to crizzle and decay, have been redressed several months ago and the glass since then have all proved durable and lasting . . .' The Glass Sellers were indeed sufficiently impressed to apply for a patent on behalf of themselves and

Ravenscroft, and a seal depicting a raven's head was chosen to identify the new and improved glass.

The new metal appeared at an apt moment in history, at a time when both the establishment and the mercantile middle class were becoming increasingly affluent, and on the threshold of an age of conspicuous consumption. The beauty and durability of the new lead crystal recommended it to those who had the purchasing power, and glasswares took their place on the tables of the well-to-do alongside the elegant silverwares of the period and the new porcelain wares that were being produced all over Europe.

The Eighteenth Century

The glasswares produced from Ravenscroft's metal during the first half of the eighteenth century were sensible, solid and elegant. Like the silverwares and the furniture of the period, they were designed to appeal to the Age of Reason, for the eighteenth century thought of itself as a new Augustan age. The forms, which derived from baroque architecture, had a classical correctness of proportion. The decoration was integral, the stems shaped into knopped balusters, sometimes enhanced by teardrops which were later elaborated into air-twists. A few glasswares were engraved, but these were the exceptions. Whereas in Germany and Holland the bowls or glasses were extravagantly engraved or enamelled, the English makers preferred to leave their bowls unadorned, believing, one supposes, that beauty of shape was beauty enough.

Cut glass

Though Ravenscroft, unknown to himself, had produced the ideal medium for cut glass, soft and highly refractive, this style of decoration only became common in England towards the end of the eighteenth century. The possibilities of this form of decoration must have been appreciated much earlier than this. The technique of diamond-cutting had been greatly improved in the seventeenth century, more accurate faceting revealing for the first time the fire and brilliance inherent in the stones. This must have suggested the possibilities of exploiting the refractivity of glass. Then, in the middle of the eighteenth century, Josef Strass, an Austrian living in Paris, began to produce paste imitations of diamond, making his pastes from a lead crystal and faceting the material in imitation of the fashionable brilliant-cut and rose-cut diamonds. These plausible imitations had considerable brilliance, and showed just what could be achieved by creating facets on glass to exploit its reflective properties. A prism shows how glass refracts the different colours of which white light is comprised to slightly different degrees, and splits this white light into the colours of the spectrum, from the reds through to the blues. Strass created what amounted to a series of prisms on the surface of his pastes, and this, too, was what the glass-cutters would eventually create on the surfaces of wine glasses, vases and decanters. All this, however, may well have been appreciated even before Strass pointed the way. Cut glass had, in fact, been on sale in England as early as 1709. In the *London Gazette* during that year a notice appeared to the effect that 'there is lately brought over a great parcel of German cut and carved glasses, viz: jellies, wine and water tumblers, beer and wine glasses with covers, and diverse other sorts'. The fact that these were not the usual engraved Continental glasswares is suggested by the statement that 'the like hath not been exposed to public sale before'. It also appears from the records that a member of the Glass Sellers, John Ackerman, brought over a German glass-cutter to England some time before 1719.

What probably restricted the production of cut glass here was the tax imposed on glass, in particular the Glass Excise Act of 1745, which levied a duty on the raw materials used by the glass-maker. This must have discouraged the making of wares suitable for cutting, for a pretty substantial glass is needed for the cutter to work on. Also, of course, when he cut glass a cutter was grinding away metal which had already borne a considerable burden of taxation.

Irish cut glass

It is significant that the Irish glasshouses, which sprung up like mushrooms during the eighteenth century, should have made a speciality of cut glass. The Waterford glasshouse, for instance, which gained an enviable reputation for its wares between 1783 and 1851, became famous for 'cut flint glass, useful and ornamental'. The products of these Irish factories were not subject to an excise duty at this time, and they could therefore be more lavish in their use of metal without pricing themselves out of the market. Cutting may also have had its attractions for newcomers to the craft, for the technique can disguise a multitude of imperfections.

It was probably these Irish wares that created the demand which eventually forced the English makers to produce cut glass despite the discouragement of the tax. Cutting in England was initially restricted to the faceting of the stems of stemwares, but by the end of the century English wine glasses with cut bowls and cut decanters were being produced in considerable volume, though the cutting at this time tended to be very restrained, and was restricted to a few shallow facets and stars. It was only after the repeal of the Excise Act in 1845 that it became feasible to produce heavy blanks into which deep mitre cuts could be made. The fact that glass now became relatively cheap to produce also led to the production of pressed imitations of cut-glass wares, which look very much like those cheap tankards to be seen in pubs today, with their all-too-apparent seams and their soft-edged facets.

The eighteenth century saw the birth of modern domestic glassware. The wine glass, still the staple product of the major glasshouses, came into general use and the forms it took are still with us. Bell, cup, trumpet, thistle and bucket bowls can still be seen in any shop selling a good selection of glasswares. Baluster stems, even air-twists (Fig. 1.6), are still to be found, and styles of cutting introduced two hundred years ago are still in fashion today. The eighteenth century also made its technical contributions to the craft, notably the lehr, the annealing furnace still to be seen in every glasshouse today. When a glass vessel has been blown and the stem and foot have been added, stresses will have been set up in the metal that would cause it to shatter if it were not annealed. In the early days this annealing took place in a chamber adjacent to the melting furnace, but it was difficult to control the temperature of this chamber and breakages were frequent. The lehr is a separate tunnel-like oven into one end of which the wares are loaded. Their temperatures are raised to the working temperature of the metal from which they are made and they then travel slowly down the tunnel, being gradually exposed to lower temperatures as they pass along. The word lehr derives from the German 'lehr ofen', meaning empty oven, and some form of primitive lehr was probably to be found in German glasshouses as

Fig. 1.6. A mid-eighteenth century English wine glass, with an air-twist stem (by courtesy of the Victoria and Albert Museum)

early as 1662. In England lehrs were not in general use before 1750, and it was an Englishman, George Ensell of Stourbridge, who, by improving the design to achieve greater heat control, produced the forerunner of the modern lehr, and made a major contribution to the industry by dramatically reducing the breakages due to those internal stresses in the glass. The eighteenth century also saw considerable improvements in the design of pots and furnaces. That today's glasswares are as near-perfect as any one-off craft-made product can be results to no small extent from the experiments carried out in the eighteenth century and to the empirical knowledge acquired by the eighteenth-century glass-makers in their search for perfection.

As was pointed out in the Preface, this book is in no way intended to provide a guide to antique glasswares. Anyone who wishes to gain a deeper knowledge of eighteenth-century glass can find any number of books that explore this fascinating subject in great detail. A browse through these volumes will reveal that the experts differentiate as many as a dozen different designs of knops on baluster stems and a comparable number of forms of foot and bowl. The reader will also discover that many types of glasswares produced at that time no longer exist, such as coaching glasses which had no foot, cordial glasses with tiny bowls and patchstands that look like miniature tazzas and which stood on a lady's dressing table to hold the patches with which she disguised her pimples. Then there were posset pots, which resembled small teapots with a handle on either side which an invalid held while he took his posset, a concoction of hot milk and spiced wine, through the spout. One could go on describing the styles and the forms of glasswares made at that time, but perhaps enough has been mentioned to indicate how, in the course of the eighteenth century, glasswares had developed from a rare and not greatly regarded commodity to being fashionable, much in demand and widely used in a variety of roles.

The nineteenth century

The end of one century and the beginning of another does not necessarily inaugurate radical changes in taste, and the styles of eighteenth-century glasswares remained popular during the early years of the nineteenth century. Gradually, however, a new spirit began to emerge, a more romantic attitude which was probably a reaction to increasing industrialization. Whatever its cause, this revolution had its effect on all the applied arts, the art of the glass-maker among them. The designers explored the lumber rooms of the past and drew upon the motifs of distant lands. To the Middle Ages, the Renaissance, India, China and Ancient Egypt, distance lent an enchantment. Designers and craftsmen also found that novelty was a saleable commodity. So when we come to look at glasswares produced during the nineteenth century we discover a bewildering variety of styles and techniques just as when we contemplate any of the other household wares of that period.

The repeal of the Excise Act meant that the resident excisemen disappeared from the glasshouse, but, more importantly, it meant that the glass-maker was no longer constrained to view his product in terms of the tax liability it would impose upon him. He now had greater freedom to create new designs. He was also more prosperous, and this enabled him to expand his operations. The repeal encouraged new firms to set up, such as Stevens and Williams, who are today one of the major companies in the Stourbridge area. Indeed, the development of Stourbridge as England's major producing area of fine glasswares dates from this period. The ready availability of coal for the furnaces and clay to make the pots attracted more and more firms to Stourbridge and beeskip-shaped glasshouses, like the one that has been preserved by Stuarts, sprung up along the banks of the canals.

Too many facets

Inevitably, perhaps, the immediate result of the repeal of the Act was a spate of cut glass, and the cutting became ever more complicated as time went on, so that by 1851 some of the cut glass on show, in the Great Exhibition held that year, was so complicated as to be quite impractical. Probably it was this excess that led to cut glass going out of fashion for a time. But another reason for this may have been the other exhibits shown at the Great Exhibition, which was held in what amounted to a giant house of glass, constructed out of 277 000 square metres (900 000 square feet) of glazed panels. Here, for the first time, the British public were shown the styles of glassware already popular on the Continent, coloured glass in a variety of hues, from deep crimson to subtle honey shades, glass of one colour overlaid with that of another, opaline glass and engraved glass.

After the Great Exhibition

In the wake of the Great Exhibition craftsmen skilled in the production of these novel and exotic wares were persuaded to emigrate to England from Bohemia and Germany, and soon the English glasshouses were turning out glasswares in the new styles. In the 1850s there was also a revival of interest in Venetian glass, and the techniques of the old Murano glass-workers were resurrected and their designs copied, or more often 'improved upon' with disastrous results. Ruskin had stated his opinion that 'all cut glass was barbarous', and the glass-makers turned again to trailed decoration, producing coloured wares, enamelling and guilding. They also made cased glassware, and revived latticinio and millefiori decoration, and they copied the Venetian ice glass by incorporating fragments of broken glass in the gather. They made 'agate glass', and produced cameo wares inspired by the greatly admired Portland Vase. The most famous producer of cameo glasswares was Emile Gallé, who, with his father, established a glasshouse in Nancy in 1867. Gallé was in the forefront of the *art nouveau* movement and was also considerably influenced by Japanese art. His cased glass consisted of metal of very subtle shades overlaid on opaque white glass. He then carved the overlay to produce characteristic *art nouveau* motifs, mainly floral themes, of great delicacy.

Wheel-engraving and intaglio-cutting

The Victorian glass-makers also decorated their glass by wheel-engraving, and intaglio-cutting enjoyed a period of great popularity. It was the development of these two techniques which was to prove perhaps the period's most important contribution to the art of glass-making, for both remain popular to this day. Indeed, intaglio-cutting is currently enjoying a considerable

Fig. 1.7. Modern intaglio-cut wine glasses (by courtesy of Stuart and Sons Ltd)

revival. Stuarts, who until recently employed only two intaglio-cutters, today have eight of them creating complex designs below small stone wheels in their main factory near Stourbridge (Fig. 1.7).

The revival of cutting

In 1880 cut glass became fashionable again and has remained so ever since. English cut crystal has retained a worldwide reputation, and by far the majority of quality glasswares produced in England are cut glasswares. Undecorated glass, relatively little of which is produced in England, has a limited appeal, though there are signs that a growing number of people, particularly the young, are showing a preference for it. This increased demand is being mainly catered for by Continental manufacturers at the top end of the market, and particularly by the Scandinavian factories. However, at the lower end of the market machine-made plain glass, produced both in England and abroad, is selling very readily. Coloured glass and similar exotic glasswares are now mainly imported from Czechoslovakia, which also, of course, produces an enormous variety of other types of glassware, including plain and cut glass.

2 Making glasswares

Raw materials

A glass-maker can produce many different kinds of glass by varying the constituents or by altering the proportions of the ingredients which he melts to produce his metal, He can make common glass, that is, soda lime glass, by fusing a mixture of sand, soda and lime. He can even produce a glass that will provide radiation protection by using a mix with a 70 per cent lead content. Then, by adding cobalt, he can make blue glass, while if he adds gold to his mix he can make glass of a rich ruby-red colour. The metal which all the makers of fine glass in the UK produce today is, however, known as lead crystal, which is particularly well suited to the production of cut glass, being both relatively soft and brilliant and also having an attractive colour.

The essential ingredients of lead crystal are silica sand, potash and red lead. Other chemicals are added, however, to facilitate the fusion of the material, improve its appearance and working qualities and ensure against discolouration. Different glasshouses may vary these ingredients slightly, but those used in the glasshouse at Royal Brierley are probably typical for the whole industry. They use:

(1) Silica sand, which contains not less than 99.6 per cent silica dioxide and which is obtained from Loch Aline on the west coast of Scotland.

(2) Lead oxide, in the form of red lead, which gives this glass its high refractive index and ensures that it has a diamond-like brilliance when it is faceted.
(3) Potash, which reduces the temperature at which the silica sand will fuse to a feasible level. This was once obtained by burning vegetable matter, but now potassium carbonate and potassium nitrate are used, as these are less likely to contain foreign matter that could contaminate the glass. The potassium nitrate, commonly known as saltpetre, has the added function of helping to bind the cullet (see below) during melting. Though soda performs the same tempering function as potash, the latter produces a far more brilliant glass.
(4) Cullet, the name given to the broken glass which is always added to the other constituents. This is obtained from the glasshouse and consists of waste from the blow irons, from the cracking-off operation and from the breaking up of rejected wares. A maker of fine glasswares will only use cullet from his own glasshouse, as otherwise he could not be sure that the ingredients conformed to the standards he laid down. He will, of course, also avoid using rejects which were discarded for faults that might be perpetuated in subsequent melts. Cullet represents 50 per cent of the total of the material in a batch and, besides the economic advantages of re-using waste, it also facilitates the chemical reaction between the other materials in the batch when they are in the furnace.
(5) Arsenic, in small quantities, because it improves the colour and the transparency and increases the brilliance of the glass. It also makes it easier to decolourize any iron impurities in the metal.
(6) Borax, introduced as a fluxing element, which also gives the glass a resistance to weathering and to thermal and mechanical shock.
(7) A small quantity of decolourizer, such as nickel, though this amounts to only about 0.27 grams in a batch of 200 kilograms, but this is sufficient to neutralize any residual iron that has not previously been removed. This is necessary because, as has been previously pointed out, as little as 0.03 per cent of iron is sufficient to give the glass a green or amber tinge.

Purification

It can thus be appreciated that modern lead crystal is a highly sophisticated material, the result of centuries of development and experience. It is vital to obtain the right materials and it is just as important to ensure that these materials are pure and uncontaminated. It is expensive to mix and fuse a batch of glass and to produce glasswares from the resulting metal. It would be disastrous if the results of all this labour had to be scrapped because the products did not conform to the high standards a modern glass factory has to maintain. So the ingredients are carefully checked and the sand to be used is subjected to a series of purification processes. This will already have been washed before it was delivered to the glasshouse, so the first step is to dry it thoroughly by passing it through a gas flame. Then the dried sand is passed through a sieve that holds back any large grains. To ensure that the constituents of the batch are evenly distributed, the grains of sand must be of approximately even size. Next, the sieved sand is passed between large magnets which remove any iron particles in it. Though these magnets are very efficient, it is still just possible for traces of iron or other metals such as chromium to escape them, which is why decolourizers have to be used. Nowadays some glasshouses obtain their sand already processed in a central plant, but, of course, this service has to be paid for.

Mixing the ingredients

The proportions of the materials used are also very critical and they are weighed out very accurately. A typical batch of 203.8 kilograms would contain 103 kilograms of processed sand, 59.4 kilograms of red lead, 32.4 kilograms of potassium carbonate, 7.7 kilograms of potassium nitrate, 0.7 per cent of borax and 0.6 per cent of arsenic, plus that tiny amount of decolourizer. English lead crystal contains 30 per cent of red lead, so conforming to the British Standard. On the Continent the lead content is normally only 24 per cent. The Czech glasshouses also produce what they call a semi-lead glass, containing only 10 per cent lead. Because it is essential to avoid any contamination of the materials and in view of the health hazard implicit in handling lead oxide and arsenic the weighing and mixing is carried out with the greatest care. The areas of the glasshouse where these operations take place are kept scrupulously clean and the operatives wear disposable overalls and face masks. The lead oxide and the arsenic are kept in sealed containers and pass to the weighing bin, and from there to the mixing drum through rubber tubes. For the further safety of the workers the atmosphere is regularly checked to ensure that the lead level is well below the danger point.

When the ingredients are all in the mixing bin this is vibrated until the batch is thoroughly mixed and is ready to load into the pots.

Making the pots

The making of the pots, in which the constituents of glass are fused, is a skilled and exacting craft, and their production represents a considerable overhead to add to the costs of buying in and processing the ingredients from which the glass is made. The clay used to make the pots at the Stevens and Williams factory comes from Rommerode in West Germany. This clay, called 'grossalmerode', produces pots capable of standing up to the considerable heat stresses imposed upon them in the furnace.

These refractory pots are built up laboriously by hand in batches of seventeen, and a batch will take between six and seven weeks to complete. The walls of these pots are 11.25 centimetres (4½ inches) thick and, having created the flat base, the pot-maker builds up these walls in courses, 15 to 20 centimetres (6 to 8 inches) at a time, using a template to ensure that the pot is true, and smoothing the clay surfaces as he progresses. The pot gradually assumes the shape of an old-fashioned bee-skip, or the nose of a space craft, and near the top a mouth is created by using a suitable mould (Fig. 2.1). When the pot has been completed it has to be left to dry out completely in a draught-free area, and this takes a further six months.

A glass furnace usually holds ten or twelve of these pots and they can only be used between twenty-seven to thirty times before they have to be discarded and replaced with new ones. This is because the chemical reaction between the clay and the metal

Making the pots   29

Fig. 2.1. Refractory pots (by courtesy of Waterford Crystal Ltd)

during firing causes pitting, and the sand content in the clay also gradually fuses to form a glass of a very inferior quality which could contaminate the melt.

The pots are filled every two days and one of them is changed every week, traditionally on a Friday afternoon. This rotation of pots ensures that there are always an adequate number in use to supply the glass-makers with metal. A broken pot is a major disaster in a glasshouse as it results in skilled craftsmen being left without any work to do. Obviously it is uneconomic to maintain a surplus of metal to ensure against all eventualities.

Replacing the pots

The new pots will have been placed in a special furnace and kept at a temperature of 1100 degrees Centigrade for three days before the time comes to place them in the furnaces in the glasshouse. They are then carefully checked to see that they have not cracked.

Now the bricks, which were used to seal in the pot to be replaced in the furnace, are broken away. Then the old pot and any waste metal that has collected around it are removed. The new pot is then manoeuvred into the space in the furnace wall and bricked in, a small hole being left to give the glass-workers access to the corresponding hole in the top of the pot.

Filling the pot and firing

The pots are filled by a team of men known as 'teasers'. They shovel in the mixed ingredients and the cullet, some of one and then some of the other, until a 50/50 mixture has been built up. The pot is now stoppered to avoid contamination and left for eight hours. By this time the level of the contents will have fallen, so the pot has to be topped up with further cullet. It is stoppered again and stays in the furnace for twenty-four hours at a maintained temperature of 1350 degrees Centigrade. The stopper is again removed and the temperature is now reduced to 1150 degrees, and this temperature is maintained for a further sixteen hours. By this time the metal will have acquired the desired hot toffee consistency and is ready to be worked.

Inside a glasshouse

The making of glasswares in a glasshouse has been likened to a strange ballet carried out in the half-gloom. The workers are in continual motion, gathering the metal glowing orange from the furnace, revolving the irons to produce the centrifugal force necessary to keep the metal evenly distributed, their faces illuminated by the glow from the furnace and from the molten metal they are carrying. Nobody who visits a glasshouse can fail to be fascinated by this strangely ritualistic performance. Nor can one fail to be impressed by the skill these craftsmen bring to their job, and by the fact that fine glasswares are still individually produced by hand, as they have been for thousands of years. Old engravings, cut centuries ago, show craftsmen employing the same simple tools and techniques that are used today. The vocabulary of the glasshouse, too, contains words that date back to the time when foreign craftsmen were brought over from the Continent to make glasswares here. The glass-making industry is a craft industry, steeped in tradition.

The chair

The glass-makers work in teams, known as 'chairs', and the work is, in fact, carried out in the proximity of a crudely made seat with extended arms, also confusingly known as the 'chair'. The number of members of the team varies, depending on the type of ware the chair is producing, but a wine chair, which produces stemwares, consists of four craftsmen. These are the servitor, the foot-blower, the bit-gatherer and the taker-in. The servitor, more often known as the gaffer, is the man in charge of the chair, and he occupies it. In a caterhole shop, where large articles such as jugs and vases are made, the man in charge is known as a 'workman'. He has a number of servitors under him who do the preliminary blowing and forming. He then shears the lip, shapes it and, where necessary, adds the handle.

Tools and techniques

The tools the members of the chair use are simple ones. There is the marver, which consists either of a block of carbon steel or a hollowed-out length of wood. This piece of equipment is as old as the craft itself, and gets its name from the fact that it was originally a block of marble. On the marver the metal gathered from the furnace is rolled, or 'marvered', to ensure that it is evenly distributed on the blowing iron.

Then there are the irons, the hollow blowing iron about 2 metres (6 feet) in length and the solid pontil or 'punty' iron. There are the various tools the gaffer uses to trim and shape the molten glass – shears, pincers and lengths of board. And there are the moulds. Nowadays only special pieces are free-blown, most standard lines being blown in a cast-iron mould. These moulds are usually mounted on a device known as a 'mechanical boy', because they replace the boy whose job it was to open and close the non-mechanical hinged mould. The mechanical boy allows the foot-blower to operate the mould himself by depressing a pedal. The interior of the mould is coated with linseed oil and sawdust, and this acts as a lubricant, allowing the foot-blower to rotate the metal in the mould while he is blowing it. Before blowing, the mould is cooled and cleaned with soapy water, and there are holes in the walls of the mould which allow this water to escape in the form of steam when the hot metal is inserted. If this did not happen the molten glass would cool too quickly, and the blown glass would crack. In a mechanical boy the mould remains immersed in soapy water until the foot-blower raises it and opens it to receive the metal.

Making a wine-glass bowl

The first step in making a stemware is the gathering, by the foot-blower, of the molten metal from the pot on the end of his blowing iron (Fig. 2.2). This may seem simple enough, but he must judge by experience how much metal to take and know how to gather it. A mistake at this stage can mean a reject later on. He

Fig. 2.2. Gathering glass from the pot

Fig. 2.3. Marvering the metal in a hollow wooden marver

inserts the blowing iron through a sillimanite ring that floats on the surface of the molten metal. Sillimanite is a silicate of aluminium, named after an American chemist, Benjamin Silliman, and it has proved the ideal material for this application. The reason it is there, floating on the molten metal, is because a scum keeps re-forming on the surface, and it is far easier to keep the area within the ring clear of this than to have to continually skim the whole surface within the pot.

The foot-blower now takes his gather of metal to the marver and marvers it (Fig. 2.3). Still rotating his iron, he now moves on to the mechanical boy, mounts it, inserts the metal in the mould and

Fig. 2.4. Blowing the bowl of a wine glass in a mould. (This foot-blower is not using a mechanical boy, and an apprentice has to open and close the mould for him)

blows it until it assumes the shape of the mould (Fig. 2.4). A foot-blower tends to develop the muscles in his cheeks, and after many years of working they can expand them to a remarkable and alarming degree to resemble balloons.

Stem and foot

The foot-blower now carries the sphere of glass he has created to the gaffer sitting in the chair. The gaffer places the blowing iron

Fig. 2.5. The bit-gatherer passing the punty iron with the gather on the end of it to the gaffer, seated in the chair

across the arms of the chair and continues to rotate it. Meanwhile the bit-gatherer is collecting a gather of metal from the furnace on his punty iron. He carries this to the chair and the gaffer up-ends the blowing iron, reaches above his head, takes the punty iron from the bit-gatherer and allows the metal from it to flow onto the base of the blown sphere of glass (Fig. 2.5) He then again places the blowing iron across the arms of the chair and, rotating it, forms the stem, using a long pair of tweezers with wooden inserts (Fig. 2.6). Watching him perform this operation, it seems incredible that he can shape these stems to the right pattern within very close limits. This operation is then repeated, but this time he shapes the gather which has flowed onto the end of the stem into a foot between two hinged boards, known as 'foot-boards'. To give the foot its final shape he uses wet, acid-free paper. In some factories large wares such as vases are produced by a combination of mechanical blowing in a mould and free blowing by a foot-blower.

Fig. 2.6. The gaffer shaping the stem of a wine glass

Cracking-off

We now have a sphere of metal on the end of the blowing iron, and the top of this sphere has to be removed to leave a hollow bowl. This is the job of the taker-in. To remove the unwanted part, known as the 'moil', the taker-in scores a line round the globe of glass where the rim of the wine glass will eventually be. This is done with either a diamond or a hardened steel tool. The glass is then passed through a gas flame. Due to the different rate of expansion in the scored area when heat is applied, the glass fractures at this point. In most factories today this operation is carried out in an automated machine, the glasses being placed on revolving chucks which brings them in contact with the diamond tool and they are then automatically transported through a flame (Fig. 2.7). A further automated machine then polishes the edges.

Fig. 2.7. Automated cracking-off

Previously, the edges were polished by a process called 'edge-melting'. The glass was mounted on a turntable and a flame directed on the rim, heating the metal to the point where it flowed and the sharp edge became softened. Cracking-off occurs at an earlier point in the production process when larger items are being produced. After blowing, a jug or vase will be immediately cracked-off and then attached to a punty iron. This allows the workman to carry out forming operations on the lip (Fig 2.8). With

Fig. 2.8. A workman in the caterhole shop shearing off excess metal from the rim of a vase

large items it is necessary to re-heat the metal, often a number of times, and this is done by placing the vessel on the end of the iron in a small furnace called a 'glory hole'.

Into the lehr

When the taker-in has cracked off the moil he then carries the glass as quickly as possible to the lehr, having first consigned the moil to a bin known as the 'shrower' to be later ground up to make cullet. The function of the lehr was described in Chapter 1. A modern lehr is some 21 metres (70 feet) in length and is heated by electricity. The glasswares travel slowly through it on a moving belt. First, however, they are raised to the working temperatures again to relieve the stresses caused by forming the metal and due to the fact that the exteriors of the wares will have cooled more quickly than their inside surfaces. Then over a period of between two and four hours, as they travel on, the temperature reduces till it reaches room temperature. Larger articles take longer to anneal; a large and heavy vase might take as long as twenty-four hours in an annealing furnace.

Acceptance and rejection

Not every glass that passes through the lehr emerges unscathed. If a glass has been worked on over too long a period it may be already too late to redress the effects of the internal strains when it goes into the lehr. It will therefore crack on the conveyor belt and the fragments will be removed by the 'taker-off'. The taker-off keeps a record of all the breakages and all the faulty blanks he removes from the lehr, and, as these can be attributed to the chair that produced them, bad workmanship can be pinpointed. An acceptable level of breakages in the lehr would be about 2 per cent. The taker-off inspects every ware individually, rejecting faulty blanks at this stage before they pass to the cutting shop. There are a number of common faults he will look for:

(1) *Stones*
Inclusions in the glass, probably due to contamination in the furnace.
(2) *Cords*
Colourless streaks in the glass, which result either from poor mixing of the constituents or from the joining of metal of different temperatures.
(3) *Blisters*
The air bubbles, sometimes known as 'seeds', resulting from the gases created during melting not being driven off. At one time these were very common, but with modern furnaces, the

temperatures of which can be more readily controlled, they occur less frequently.

(4) *Speckles*
Caused by dirty tools contaminating the metal.

(5) *Bad shearing*
The result of shearing the excess metal from the rims of jugs and vases, etc. either carelessly or when the metal is too cold.

(6) *Mould Marks*
Caused by the metal walls of the mould having become damaged, or more frequently by the failure of the foot-blower to rotate the metal sufficiently during blowing.

(7) *Crizzle*
Even today, crizzling, once the bane of the glass-maker, does occasionally occur, with the surface of the glass exhibiting characteristic cracks.

(8) *Surface imperfections*
These usually have a wavy appearance.

(9) *Uneven distribution*
The result of lack of care in handling the metal. The constant rotating of the irons is designed to prevent this happening.

(10) *Glasses of insufficient weight*
All the blanks are weighed and only a small tolerance is allowed. At Royal Stuart, for instance, a variation of not more than 7 grams (a quarter of an ounce) is permissible for stemwares, though for very small glasses such as liqueurs the tolerance would be smaller.

(11) *Glasses which are too tall or too short*
All blanks are checked for height and only very minor variations are permitted.

(12) *Lopsided wares*
These are called 'sidens', and again result from insufficient care being taken in handling the metal.

(13) *Other faults*
Among these are included glasses with a foot or bowl which is oval rather than round, with feet which are set at an angle (known as 'wob foot'), with badly made stems, with chipped feet or rims and with rings on their feet caused by a gaffer using worn foot-boards.

One piece of apparatus which is invaluable to the taker-off is a turntable. If a ware placed on this is rotated it will, if it is a siden, flash as it revolves, and so it is easy to see if a ware is out of true. For this reason sidens are sometimes referred to as 'flashers'. Another reason why glasswares may be rejected is because of their colour. This may vary very slightly from batch to batch, and though by carefully matching suites this can usually be coped with, any considerable variation in colour would result in rejection.

The rate of rejection is high, and this is, of course, reflected in the price of fine glasswares. The rate from the lehr and inspection after cutting would probably run at somewhere about 30 per cent of all the wares produced. If a factory makes plain glasswares, or glasswares with plain bowls such as Stuart's air-twist pattern, the rejection rate can be as high as 50 per cent.

Of the rejected glasswares coming out of the lehr some will be consigned to the shrower, the bin which holds the glass that will be converted to the cullet. But wares exhibiting such faults as speckles, stones and chords or crizzling would be completely scrapped, because to re-use them would be to perpetuate the problems. Some faults can be removed in the course of cutting, however, and the taker-out will use his experience and judgement in deciding what can be salvaged in this way. Any glass-buyer would be well advised to compile a checklist of common faults similar to that used by the taker-out, so that he can check his deliveries. The standards of inspection of a new supplier might not be as high as the firms he has dealt with in the past, and even established suppliers sometimes let wares through that should have been sent to the seconds shop. One would think, too, that a turntable was an indispensable tool to a glass-buyer's office, as it is not always easy to detect sidens just by looking at them.

Machine-made glasswares

It is becoming increasingly difficult to distinguish machine-made glass from the products of a traditional glasshouse. In the past a sure sign of a machine-made glass was a rolled lip as compared with the straight-edged lip of a blown glass. Nowadays, however, mass production techniques have greatly improved. Also, many machine-made wares now have their bowls blown in batches in sophisticated machines, and do not therefore have that characteristic lip. Also, modern moulded imitations of cut glass do not always exhibit that seam which was once an immediate give-away, because many machine-made wares are now produced in one piece. The modern machinery used in the making of mass-produced glasswares can be illustrated by a brief description of the plant at the Thomas glass-works in West Germany. The metal is continuously smelted in an open furnace and is fed by a continuous process into the sixteen injection moulds which form the bowls. The metal is first pre-shaped and then blown and simultaneously turned in the moulds by compressed air. Then the stems and feet are formed in sixteen presses and in a second mould are assembled onto the bowls under pressure. After annealing, further automated machines crack the moils and polish the rims.

With experience one can, of course, learn to tell if a glass is hand- or machine-made. It is really a matter of looking at it. But one guide is that machine-made glasses will often exhibit one or more of those faults which would have led to their being rejected as they came out of the lehr had they been craft-made. No manufacturer of machine-made glass could accept the high rejection rate which a manufacturer of cut crystal must have if he is to maintain his reputation. Taking a dozen glasses at random from a well-known maker of mass-produced glass, I discovered either cords or blisters in every one of them.

3 Decorating glass

Glass can be decorated in a hundred and one different ways. This decoration may be integral, that is, incorporated in the glass in the process of forming the metal into a ware. Effects such as latticinio and trailed decoration used to create decorative stems are included in this category. Then there are the various ways in which decoration can be added to glass after a blank has been produced. Either the glass may be cut away, or decoration such as gilding or enamelling may be applied to it (Fig.3.1).

Glass is, of course, an inherently beautiful material, and there are those who consider that any form of decoration is extraneous, that the line and form which the designer dictates are beauties enough. Indeed, when one looks at some of the graceful products of the leading Scandinavian glasshouses it is not difficult to understand why some people feel that this is glassware at its best, that to decorate such shapes is to pile Pelion on Ossa. Similarly, eighteenth-century wine glasses, the decoration of which was limited to a teardrop or an air-twist in the stem, are eminently satisfying to look at and pleasant to use. But whether one agrees with Ruskin that to cut glass is a form of sacrilege, the fact remains that the majority of glasswares sold in the UK today are decorated, and by far the most popular form of decoration is cutting. However, before looking at this and other popular forms of applied decoration, let us see what the glass-maker in the glasshouse can achieve in the course of working the metal, and what can be achieved by varying the constituents of that metal. For although these techniques are not widely used by the larger British glasshouses in the production of table glass, they are still employed

Fig. 3.1. Painting flowers in enamel on the opaque white surface of cased glass in a Bohemian glass factory. Afterwards the enamel is fired

by smaller firms and designer-craftsmen in the creation of decorative glasswares in the UK and on the Continent, particularly in Czechoslovakia.

Early glasswares, it will be recalled, were invariably coloured. Initially this was because the glass-makers did not know how to make colourless glass. Later, after this technique was well known, coloured and opaque glass continued to be produced. Long after the Venetians had demonstrated how to produce cristallo and this knowledge had been disseminated throughout Europe, the German glass-makers continued to produce their green and golden tinted waldglas. During the intervening centuries coloured glass enjoyed periods of popularity, and colour was the most popular form of integral decoration. Currently it has only a limited sale, but there are at least some early signs of a revival of public interest in it.

There is, of course, a world of difference between the accidently coloured glasswares of the past and the coloured glasswares produced during the past two centuries or so. From the earliest period of glass-making the manufacturers seem to have discovered how to induce colour in glass, as the trailed decoration on early

Egyptian glass vessels indicates. However, it is almost certain that these craftsmen did not have much control over the coloured glass they produced, and the tint of their blue trails was no doubt fortuitous rather than calculated. Today, we can, of course, produce a wide range of coloured metals of predictable hues. Blue glass, which was the colour favoured by the early glass-makers, is produced by the addition of a small quantity, about 0.05 per cent, of either cobalt oxide or copper oxide in a batch. Opaque white glass, which also has a long history, was, it will be recalled, used as an overlay on the Portland Vase. This is produced by using a calcium antimony compound. A fine ruby-red glass can be produced by adding gold or copper, while the latter is used in the making of red and green glass as well as blue glass. By experimenting with various metallic oxides the Victorian glass-makers were also able to produce glasses of subtle hues and, if the demand existed, any modern glasshouse could provide a wide range of coloured glasses. Some idea of the possibilities can be gained from looking through the lavish publications produced by the government-owned Czechoslovakian glass industry. Not only do the Czech glasshouses and the artist-craftsmen there make glass in many subtle shades, they also produce decorative wares by overlaying two or more different coloured glasses to produce what is called cased glass.

Glass on glass

One only sees prunts today on reproduction glasswares made in West Germany for the tourist trade. Another seldom-used technique is trailed decoration, though one does occasionally come across it applied to decorative wares. The technique of producing trails is very similar to that used for those little glass figurines and glass animals that were so popular a few years ago. A coloured glass rod is melted in a gas flame to working temperature and then formed to the desired shape with simple tools.

Balusters and twists

The baluster stem is probably the most satisfying example of integral decoration ever produced by a glass-maker. It is functional as well as decorative, and its subtle curves exploit the beauty of the material. It was borrowed from the baroque balustrade and was first applied to glass in the seventeenth century, but it is seen at its best in the wine glasses made in England during the eighteenth century, a perfect foil to the simple undecorated bowls. It is still very much with us today. Probably half of all the stemware patterns produced in the major glasshouses have baluster stems, though these are now, more often than not, faceted. Some, however, like Stuart's Glengary pattern glasses, still have undecorated balusters, and very effective they are. The gaffer has to create freehand the knops that make up these stems, and perhaps because we are nowadays so concerned that every glass in a suite should match every other as closely as possible, today's baluster stems tend to be a lot less complex than many of those produced in the eighteenth century. Eighteenth-century balusters often had a teardrop enclosed in them, a charming detail that one would like to see more widely used today. The first tears were probably the result of an accident, the trapping of an air bubble when the gather of metal from the bit-gatherer's punty iron was attaching itself to the base of the blown bowl. However, the glass-makers soon learnt to induce these tears by deliberately pegging the metal, pushing a metal spike into the gather and then covering the hole they had created with further metal. As the metal was drawn out and shaped, the air bubble so created assumed the characteristic form. Again it may well have been pure chance that directed the gaffer's attention to the possibility of further drawing out the bubble and twisting it to create what amounted to a spiral tube running down inside the length of the stem. These so called 'air-twists' were first produced in the 1730s, and by the 1760s the technique was widely used. Though one does see air-twists used to decorate baluster stems, they look far better in a straight stem, and most of the air-twists made in the second half of the eighteenth century were of this type. Air-twists are not common in modern stemwares, but they do exist. For many years now, for instance, Stuart have produced an air-twist pattern appropriately called Ariel and the Czech glass-makers also produce them.

The earliest air-twists were produced by elongating a single air-bubble, but the designs were soon elaborated by pegging a number of holes in the metal, covering these, and then twisting the metal as it was drawn out to produce the stem. Today Stuart's gaffers have a special tool with four spikes which does the pegging in a single operation.

The next development was to produce opaque and coloured twists, though they never seem to have enjoyed the same popularity as the air-twist. These were produced by a quite different technique, a number of short opaque white or coloured glass rods being incorporated in the gather before this was twisted and drawn out to create the stem.

Glass-makers have evolved many other ways of decorating glass in the course of working it. These include dipping the molten glass in water and then reheating it to produce an ice-like effect and incorporating threads of opaque glass in clear glass to create lace patterns, but these types of decoration have only a limited appeal as novelties. A more popular form of decoration is to create a fluted pattern by blowing the glass in special moulds. Machine-made glass, of course, always has the decoration imparted to it in the mould.

Cutting and engraving

By far the most popular form of decoration over the past two hundred years has been that produced by cutting the surface with diamonds or abrasive wheels. The most familiar of these forms of cut decoration is what we call glass-cutting. The first operation to be carried out on blanks is to mark them up. This work is usually done by women, and they first paint grids on the wares, one woman painting on the horizontal lines and another the vertical ones. The paint, a mixture of cirium oxide and resin, is applied with a brush and the form which the grid takes is dictated by the pattern which it is intended to cut into the ware. Figures 3.2 and 3.3 show how this marking of the vertical and the horizontal lines is accomplished. The woman who paints on the horizontal lines has her work guided by the perforated drum below the turntable on which the ware to be marked is placed, the drum being set up in accordance with the type of ware to be marked and the pattern. The rather strange pantographic device used by the woman

48   Decorating glass

Fig. 3.2. Painting the vertical lines of the grid to guide the cutter

painting the vertical lines ensures that these are spaced at the correct intervals round the ware.

These lines are, it must be emphasized, no more than guidelines within which the cutter will work. The fact that he has some latitude allows him to remove minor flaws in the blank in the course of making his cuts. The more complicated the pattern he cuts on a ware, the easier it is for him to eliminate these blemishes. This explains the seeming contradiction that the simpler the design of a cut-glass ware the higher priced it is likely to be. If the cutting is restricted to a few simple facets round the base of a stemware bowl then a perfect blank has to be used, and perhaps only 40 to 50 per cent of blanks coming out of the lehr would be suitable. If a very elaborate design is cut on to the blank, then perhaps 20 to 25 per cent of the blanks could be used. So it pays a glasshouse to

Cutting and engraving 49

Fig. 3.3. Painting the horizontal lines

concentrate its bulk production on the more elaborate patterns in its range.

The glass-cutter sits at an electrically powered lathe, which turns a carborundum or sandstone grinding wheel. He holds the ware above the wheel and, looking through it, he follows the guidelines, first making those diagonal facets dictated by the design. Then he makes the necessary straight cuts to fill out the pattern. Water pours continually over the wheel from a small pipe placed above it which washes away the glass dust produced by the cutting and keeps the work cool. If the glass is not kept cool then 'firing' may occur, resulting in uneven cutting or, even worse, in the surface of the glass flaking off.

Dressing the wheels

The cutting is carried out in two stages. First comes the roughing, which is done over a carborundum wheel (Fig. 3.4). These used by the rougher, are shaped or 'dressed' to the profile he requires for the particular type of cut he intends to produce. This shaping is done either with a diamond or a serrated metal tool which is held against the wheel as it rotates in the lathe. To produce diamonds, splits or stars the rougher will produce what is known as a 'mitre carbo', while to create scalloped-shaped cuts such as those one often sees on the neck of a decanter and to facet wine-glass stems he will produce what is called a 'fluting carbo'.

Fig. 3.4. Rough-cutting the glass on a carborundum wheel (by courtesy of Stevens and Williams (Royal Brierley))

After being used to cut two or three pieces the wheels have to be re-roughed, or 'tinned-up' as it is called, to re-open the grain so that they will cut efficiently. Then, after cutting some twenty wares, the wheels will have to be dressed all over again. The life of one of these wheels is only about two weeks, after which it will have worn down to the point where it can no longer be used. So it can be appreciated that in a factory employing a hundred cutters the replacement of the carborundum wheels is a major expense.

The rough and the smooth

When the rough cutting has been completed the facets that have been created will have a rough texture. They will be white in colour and will lack sharp definition. The next process is to smooth these facets on the brown sandstone wheels. These sharpen them up, smooth the surfaces to a sleek transparent grey colour and remove any of the carborundum that may have been left on the glass during the roughing process. Again, these wheels need constant attention, being tinned up at regular intervals with carborundum to keep the grain of the stone open. The lips of jugs and vases, what are known as 'fancies', then have to be bevelled on a bevelling stone and afterwards polished smooth, or 'linished' on a continuous emery paper belt.

Mechanized cutting

In some factories the cutting of glass has been mechanized, the wares being cut by machines in batches. To date, the designs which can be produced in this way are very limited. In time, no doubt, this drawback will be overcome and mechanized cutting will be more generally adopted. At present, though, almost all English cut crystal is still cut by hand.

Polishing

Though the cuts are now smooth, they still have to be polished. First they are 'wooded', that is, polished with pumice applied to a wheel made from popular wood. They are then acid-polished. This is done by first washing them thoroughly and then dipping the glass in a bath of hydrofluoric and sulphuric acids. In a modern factory the washing is carried out in an ultrasonic bath containing a solvent, which removes any grease that may have got on to the glass and which would inhibit the action of the acid. Fancies are still, for the most part, polished by hand, the operative wearing protective clothing and heavy-duty rubber gloves, swirling the ware round in the acid to ensure an even polish. In many factories nowadays the acid-polishing of stemwares, tumblers and other standard lines has been automated. The wares are stacked in rotating barrels which carry them through the acid bath. The two acids used for the acid-polishing each have a separate function. The hydrofluoric acid attacks the surface of the glass and the sulphuric acid acts as a cleansing agent. The timing of the operation is, of course, critical.

Final inspection

The wares, having been thoroughly cleaned, are now ready for another inspection. The inspector will be on the lookout for any faults in the blank which have not been eliminated in the cutting. Some of these could still be rectified. Blisters can often be eliminated by a skilled 'smoother', and pieces exhibiting these will be returned to the cutting room. Faults which cannot be remedied condemn the wares to be sold off as seconds. The inspector will also be checking that the cutting is even, smooth and sharp, and that there is no chipping. He will also want to make sure that none of the cuts has gone too deep, piercing right through the glass. Any glassware which fails to live up to the high standards set by the firm is either rated a second or scrapped altogether. Nowadays most of the major glasshouses have a seconds shop on the premises, and the fact that these receive a steady supply of stock gives some idea of the high standards that these firms insist upon. It also gives a good indication of the difficulty of producing wares of a quality which a firm such as Royal Brierley or Stuart is prepared to put its name to and its reputation behind. Skilled craftsmanship produced by expensive training programmes, modern equipment and modern technology can only succeed in ensuring that some 70 per cent of all the wares produced will pass inspection. And, of course, despite the rigidity of the inspection, human errors do occur, and a small number of pieces get through to the customer that should have gone to the seconds shop or into the shrower. The proprietor of a retail shop or buyer of glassware should therefore not take his deliveries for granted, even if they come from a firm of the highest repute. Of course, a reputable firm will stand behind its product and replace it, but customer complaints are the last thing a retailer wants. He should consider himself as the last inspector, right at the end of the production line, and establish his own standards. It must be appreciated, however, that hand-made glasswares will inevitably vary slightly from one another. Only a machine can reproduce a design perfect in every detail time after time. It is not uncommon for customers to come back to a shop from which they have bought a suite of stemwares and complain that they are not all the same. This complaint is sufficiently common to make it essential for those who work in a shop selling glasswares to know enough about the making of fine lead crystal to be able to satisfy these customers. Indeed, if everyone who sells and everyone who buys glass could spend a day going round a glasshouse, not only would such problems not arise, but there would be a wider appreciation of the fact that fine glasswares are not expensive, but surprisingly inexpensive when one sees just what is involved in producing them.

Intaglio-cutting

A typical intaglio-cut glassware is illustrated in (Fig. 3.5). It can be seen that the cutting resembles that which is applied to seal stone. Compared with the mathematical cutting of cut glass wares, it is far more naturalistic, the curved forms of stylized leaves and flowers being represented in a realistic way. This form of decoration has been out of fashion for some years, but it is now enjoying a revival. Some measure of this is the fact that at the Stuart factory outside Stourbridge they have recently increased the number of intaglio-cutters they employ from two to eight. The intaglio-cutter also works with a sandstone wheel turned by an electric lathe, but his wheels are smaller than those used by the cutters of cut glass. He also works under the wheel and not over it (Fig. 3.6), and this allows him the freedom to create the curved cuts which are characteristic of intaglio decoration.

The designs which the intaglio-cutter produces may be acid-polished like cut glass, but often they are left with the frosted finish which they have when they come from the wheel. Sometimes part of the design will be polished while other parts of it are left frosted by stopping them off before the ware goes into the acid bath. Stuarts have, for instance, recently produced a new suite of stemware which has a faceted stem with a frosted finish and a bowl intaglio-cut with a flower pattern which has been acid-polished. This is very effective, and one would hope that this

Fig. 3.5. Intaglio-cut glassware from Czechoslovakia

Cutting and engraving 55

Fig. 3.6. Intaglio-cutting below the wheel

combination of faceting and intaglio-cutting with contrasting finishes will be further exploited in the years ahead.

Engraving

Another method of decorating glass by cutting away the metal is engraving, of which there are two methods. The metal can be scribed with a diamond (Fig. 3.7), and this is a technique usually employed by individual artists in their own studios working on blanks supplied by the glasshouses. The special pieces which these engravers create are often highly desirable works of art produced on a one-off basis and are, of course, extremely expensive. The other technique of engraving glass is known as wheel-engraving. This is carried out under a copper wheel of small diameter which is charged with a mixture of oil to prevent 'firing' and emery dust,

Fig. 3.7. Diamond-engraving Royal Brierley's inaugural chalice for President Reagan (by courtesy of David Whyman)

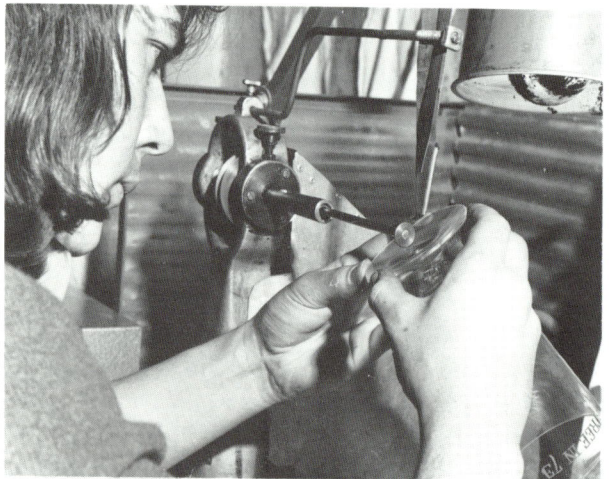

Fig. 3.8. Copper-wheel engraving (by courtesy of the Wedgwood Group)

which acts as the cutting agent (Fig. 3.8). Though wheel-engraving is less laborious than diamond engraving, it nevertheless employs many hours of a highly skilled craftsman's time, and the technique is therefore generally only used for commissioned pieces. It is often difficult to tell whether a design has been created by a wheel-engraver or an intaglio-cutter, and it has indeed proved possible to reproduce some designs, that would have previously been engraved, more economically under the intaglio wheel.

Sand-blasting

Because of the high cost of either intaglio-cutting or engraving it is now common practice to decorate popular crested wares by sand-blasting. The result closely resembles engraving, but the practised eye can readily detect which technique has been used to

Fig. 3.9. Sand-blasted decoration added to a cut-glass goblet (by courtesy of Stuart and Sons Ltd)

add these embellishments. Nevertheless, sand-blasting does produce a very acceptable result at a price which makes such pieces available to a far wider cross-section of the public (Fig. 3.9).

The glasswares which are to be sand-blasted first have a silk-screen printed mask fitted around them. This consists of tissue-paper printed with a rubberized ink which covers the areas which are to be left unscored (Fig. 3.10). The ware is now placed

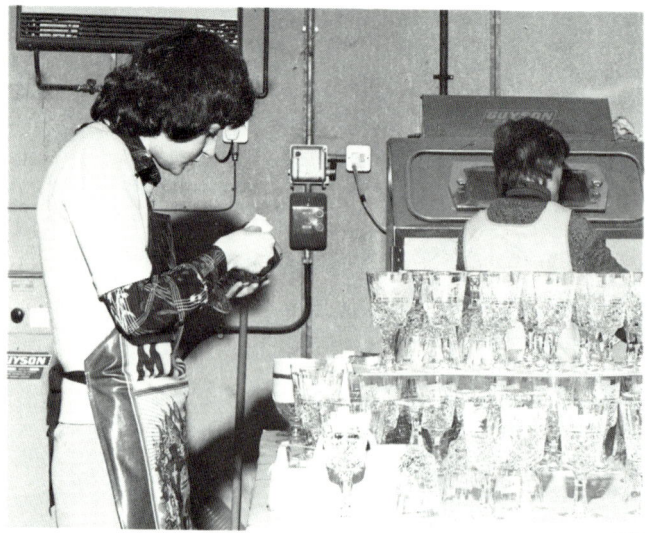

Fig. 3.10. Glasswares being decorated by sand-blasting. The woman in the foreground is applying the rubberized masking

on a turntable in a unit with a plate-glass lid. Operatives place their hands inside the built-in heavy-duty rubber gloves, activate the jet of abrasive powder and move the glassware on its turntable to ensure that the areas of glass unprotected by the rubber masking are evenly abraded. The job is far from being unskilled, but it produces the design in a fraction of the time it would take on a wheel.

Price justification

From what has been written in this and Chapter 2 it is not hard to understand why a suite of cut crystal stemwares now costs many hundreds of pounds. The cost of the raw materials, red lead and potassium carbonate, is not inconsiderable. Then there is the expense of ensuring that the raw materials are uncontaminated, of mixing them in exact proportions and of ensuring the safety of the operatives. The cost of labour in the glasshouse and in the cutting department is very high, because not only is glass-making very labour-intensive, but the craftsmen are highly skilled and have to be paid accordingly. To maintain standards, new recruits have to have expensive training before they can become productive. Another factor which contributes to costs is, of course, the high level of rejects, and though some of these can be salvaged, to produce a very acceptable cash flow from the factory's reject shop, they are nevertheless not as profitable as the wares that pass the final inspection and are sold at the full price. One can, of course, justify the price of fine crystal on aesthetic as well as on economic grounds. These wares have a beauty all of their own, and though no-one would deny that mechanically produced glasswares are today far more attractive that they were in the past, they still do not have the look or the feel of the craft-made product.

4 The discovery and development of the potter's craft

> China's the passion of her soul;
> A cup, a plate, a dish, a bowl,
> Can kindle wishes in her breast
> Inflame with joy or break her rest.
>
> *John Gay* (1725)

When man found that he could change malleable and impermanent clay into that all but indestructable substance which we call pottery he took an important step along the road to civilization. Civilization resulted from the ability of tribal communities to obtain a surplus of food in excess of the daily requirements of the members of the tribe. No longer did they have to wander to gather food and hunt in order to subsist. But the existence of a surplus implied the need to store the precious grain, oil and water which were the life-blood of the community, and pottery jars proved to be ideal containers for both solids and liquids.

Archaeological evidence has revealed that man first made pottery in the Neolithic period, in about 6500 BC, in the Near East. Over the centuries this useful art, the art of the potter, was widely disseminated, and the sites of ancient villages, towns and cities are littered with pottery fragments, the potsherds of the archaeologists. Some very early pottery has even survived intact for thousands of years, a tribute to the resilience of fired clay. Those who have visited Crete can hardly have failed to be impressed by the huge storage jars at Knossos and Phaestos, relics of the Minoan civilization that flourished there in the second millennium BC. A few years ago, it may be recalled, Jacques Cousteau found on the bed of the Mediterranean the wreck of a trading vesssel which had been loaded with hundreds of amphorae, again reminding us how important pottery storage vessels were to the peoples of the ancient world.

At an early date, too, man discovered other uses for pottery. He fashioned cooking pots from it, having found that cooking rendered the food he stored more palatable. He also made dishes and cups from which to eat his food and modelled images of the fertility goddesses, so creating the forerunners of today's pottery figurines and of the votive figures still made in some parts of the world. Just how important pottery was to those early agricultural communities can be gauged from the fact that until quite recently there were no fewer than five million village potters working in India where life has changed little over the centuries.

The beginnings of science

V. Gordon Childe, in his famous book *Man Makes Himself*, has suggested that the discovery of how to make pottery had a significance far greater for the development of civilized living than the solving of the storage problem. Pot-making, he writes, was perhaps the earliest utilization by man of a chemical change. The essence of the process is the expulsion by heat of some molecules of water from the hydrated silicate of aluminium which is the chemist's name for potter's clay . . . There was more to this than might be thought, Childe pointed out. To mould clay it must be wet, 'but if you put your damp pot straight into the fire it will crack', and even in its crudest form the potter's craft is complex. He also made the point that building up a pot was a supreme instance of creation by man. The artistic element in pottery has indeed remained important from the earliest times right down to the present day.

The potter as artist

From the beginning, potters in different areas began to create their own distinctive styles of pottery, and as time went on and tribes

Fig. 4.1(a). The coiling technique. Here, Valerie Calver of the Trident Pottery has formed a circular base and is rolling out clay into long sausages

Fig. 4.1(b). The rolls of clay are then coiled, the joins between layers are smoothed with the hands and a pot gradually built up

invaded one another's territories or trade introduced new ideas, styles changed, new forms appeared, new types of decoration were evolved. So distinctive were these different styles of pottery that, prior to the development of carbon-dating, the potsherds found in the layers of ancient sites were invaluable to the archaeologist in helping him to date other finds and also in charting the movement of peoples before the existence of written records. As they moved, people took their pottery styles with them, so that one finds the pottery of the conqueror lying on top of that made by the conquered. We even identify some ancient peoples by the pottery they have left behind them; among these are the 'corded ware people' and the 'beaker people' of western Europe.

Pottery, too, came to be an important article of trade. The Minoans in Crete, who raised the making of pottery to an art, were able to export it not only to Mycenaean Greece but also to Egypt, where they probably bartered it for gold and grain.

Before the wheel

To begin with, the potter had no wheel on which to throw his pots. The potter's wheel was not invented until 1500 BC, when it was known to the Ancient Egyptians, and did not come into general use until even later. In early times, therefore, pots had to be built up laboriously by the coiling technique, sausages of clay being wound round and round one on top of another (Fig. 4.1). In recent years a number of studio potters have demonstrated that it is possible to make workmanlike and shapely pots by this technique, but the introduction of the wheel both speeded up the process and made it easier to create elegant shapes.

The beginning of glazing

The earliest pottery was, of course, unglazed earthenware and therefore, to a degree, porous, like the plant pots we use today. The earliest glazed wares were almost certainly the Egyptian faience wares made about 4000 BC. These were, however, more akin to glass than pottery, consisting of powdered quartz suspended in a glass matrix which was then covered with a further layer of glass. The technique of applying a glaze over earthenware to render it impervious to liquids probably only became widespread in the Middle East in the first millennium BC. It was at this period that coloured glazes were used to produce those beautiful brick reliefs on the walls of Nineveh and Babylon and in Darius's palace at Susa.

Meanwhile the civilizations that succeeded one another in the countries located round the eastern Mediterranean continued to develop characteristic styles of pottery decoration (Fig. 4.2). The

Fig. 4.2. A Mycenaean drinking cup from the thirteenth century BC (by courtesy of the British Museum)

painted pots of Minoan Crete, decorated with bulls and writhing octopi, were succeeded by the creations of the early Greek potters, with their geometric decoration. This gave way after 675 BC to the Corinthian black figurewares and to the red figurewares of Athens, pottery that has never been surpassed for the artistry of

Fig. 4.3. A Greek red-figured *skyphos*, from about 490–480 BC (by courtesy of the British Museum)

its decoration and which gives us such an insight into the way of life of the ancient Greeks (Fig. 4.3).

Colour

The colouration of the earliest pottery had resulted from the impurities present in the clay the potter had at his disposal and the material he used to fire his pots. Decoration was initially restricted to designs incised into the clay or, as in the case of the corded wares, applied by pushing some object into the clay before firing, in this instance a length of cord. As time went on, however, not only did the potters succeed in controlling the colour of the bodies, but also, as we have seen, the decoration they used became increasingly sophisticated. Some pottery was decorated with slip coloured by metallic oxides, coloured glazes were applied to it or designs were painted on the surface of the pottery in enamels which were then rendered permanent by firing. In fact, many of the techniques which the twentieth-century potter makes use of had already been evolved in the Middle East and in eastern Europe over 2000 years ago.

The Chinese contribution

It was in China, however, that the art of the potter was to be raised to new heights, and it was also here that two most important technical innovations were to be introduced. At first, the Chinese potters, like those elsewhere, produced earthenware storage vessels and vessels for cooking, eating and drinking. The wares they made in the fourth and fifth millennia BC, some of which can be seen in the British Museum, are not greatly dissimilar from those made in the Middle East at the same period. For a time, in fact, the potters of Persia became technically more accomplished than their Chinese counterparts. But the Chinese had a natural resource in the form of large deposits of kaolinite, a pure white clay, which, when the Chinese potters eventually learned to exploit it, was to make their pottery famous throughout the world.

At first, any difference in the body produced by one potter as distinct from another was purely fortuitous, the result of the clay which happened to be available and the potter's ability to operate his kiln. The Chinese changed all that. They began to produce different types of pottery by design, and as a result the term 'pottery' became a generic one for a number of substances, the chemical and physical characteristics of which were quite different.

At an early date the Chinese differentiated between low-fired wares, which we know as earthenwares and which they called *t'ao*, and high-fired wares, which we call stonewares and which they called *tz'u*. They were already making stonewares by the sixth century AD, having learned to build furnaces which provided the high temperature needed to fuse the constituents of the body into an impervious stone-like substance.

Glost from the kiln

In the Shang period (about 1600 BC) the Chinese potters had already come to appreciate the advantages of making their pots impervious to liquids, and had developed glazes which enabled them to do this. A glaze is a glassy substance, as the name implies, which is fused onto the pot to provide a sleek outer skin. The earliest glazes were probably produced as a result of accident. Probably ash containing silica got blown onto the pots in the furnace and was fused to produce shiny areas tinted green, yellow or brown. This often provided a pot with gratuitous but attractive decoration, and the Korean potters began to deliberately encourage this happy accident by stirring their kiln fires. This furnace ash was known as glost, and to this day the kilns in which glaze is fired are known as glost kilns.

By the Shang period, however, the Chinese potters had begun to produce glazes on a much less hit-and-miss basis. They began to apply them to their wares before putting them into the glost furnace. We do not know what the earliest glazes were composed of, but by 1000 BC we know that lead glazes were being used. As lead is a good medium for applying oxide or iron, copper and manganese, we can assume that this is the secret of the beautiful coloured glazes we find on the Chinese pottery of this period.

The discovery of porcelain

Gradually, as the centuries went by, and the Chinese potters came to better understand the chemistry of glazes and bodies and to improve the design of their furnaces, so their pottery became ever more sophisticated. Then in the Tang period, during the ninth and tenth centuries AD, a development took place that was to revolutionize pottery. It was as a result of the search that had gone on for centuries to discover a way of making pure white stoneware that the translucent material was eventually produced which was to become known as porcelain. In 1941 kilns were discovered near Ting-hsien in China, an area once famous for its translucent white wares covered with an ivory glaze, and these kilns were probably where the first porcelain was fired some 1100 years ago.

Porcelain is made from two principal ingredients. One of these is a white plastic clay consisting largely of kaolin, which is found in many parts of China. This by itself is useless to the potter, because it would need to be fired at the impossibly high temperature of 1750 degrees Centigrade. So a second ingredient is needed to temper the kaolin. The Chinese used a feldspathic stone known as China stone. Though old texts suggest that these are the only ingredients on Chinese porcelain, in the West a third ingredient is added, quartz, the function of which will be described later in this book.

New forms of decoration

Before porcelain was produced the Chinese used to decorate their pottery by laying beautifully coloured glazes over carved designs or designs impressed into the clay from a mould. The existence of the new translucent white clay must have suggested the need for a new form of decoration. Having gone to so much trouble to produce a pure white body it obviously did not make sense to cover it with a coloured glaze. It was indeed only logical to further emphasize the whiteness by applying to it a white glaze that would also give the wares a translucency. The white pottery that came from the kilns of Chien-tzu ts'un and Yen-shen Ts'un at this period, beautiful as they were, were perhaps too austere for many tastes. It was not, though, until two centuries had passed since the first porcelains had made their appearance that the type of decoration which for centuries was to be the hallmark of Chinese porcelain was first employed.

Blue below the glaze

It was in the thirteenth century that the Chinese evolved the art of painting in blue under the glaze, arguably the most effective type of decoration ever applied to porcelain (Fig. 4.4). It was the special physical characteristics of cobalt that made this decoration possible. As early as the Tang dynasty the Chinese potters had used cobalt, imported from Persia, to colour their glazes. The Persian potters, who had developed their own very distinctive styles of pottery by this time, had used cobalt blue to paint their pottery before the thirteenth century, but only with limited success, as the colour tended to run probably because of the type of body they applied it to. Now, however, the Chinese succeeded where the Persians had failed. The Tang potters began to paint motifs, mainly derived from Chinese mythology, on their pots

Fig. 4.4. A Chinese porcelain dish, painted in under-glaze blue, from the mid-fourteenth century (by courtesy of the Victoria and Albert Museum)

after the biscuit-firing of their porcelains. The cobalt blue was fused with glass to produce a frit. This was then ground down into a powder, suspended in oil, painted onto the pottery with a brush and subsequently fired. Then the glaze was applied and fired, covering the painted surface with a transparent, protective skin. The beautiful pots which resulted were soon in demand throughout the Middle East, and China built up a very substantial export business on the basis of these blue and white wares.

The importance of the blue oxide of cobalt was that, unlike many metallic oxides, it could withstand the high temperatures (1280 degrees Centigrade or higher) required to fire the feldspathic glazes used on porcelains.

At this period in western Europe the pottery that was being produced was very crude in comparison with the Chinese wares. There was a world of difference between the lead-glazed earthenwares and slipwares produced by the medieval European potters and the green celadon wares of the Chinese potters, or the creamy porcelains, with their intriguing designs in blue. Even the more attractive Spanish maiolica wares, the Italian tin-glazed wares and the faience wares that were developed by the European potters over the centuries lacked the sophistication of the Chinese porcelains. When these porcelains began to be imported in increasing numbers in the seventeenth century in the course of trade with the East, they were universally admired. They therefore represented a challenge to the Europeans.

Hard-paste porcelain

In search of the formula

The Dutch East India Company was responsible for importing much of the Chinese pottery that found its way to Europe at this time, and the potters of Delft were the first of the Europeans to try to copy the Chinese blue and white patterns. Their imitations were made from tin-glazed earthenware, and these Delftwares, which were sometimes quite plausible copies of the Chinese originals, are sometimes known as maiolica ware or faience. It was not, however, until the beginning of the eighteenth century that a European pottery began to produce the hard-paste porcelains that the Chinese had evolved four centuries earlier. As early as 1675 the German physicist Ehrenfried Walther von Tschirnhaus began to experiment with minerals to try to discover the secret of the Chinese. By 1694 he had made some progress, and despatched a sample of the body he had produced to his friend Leibnitz. Then in 1702 he met the alchemist Johann Friedrich Böttger, and was presumably able to persuade him that it might be more profitable to produce porcelain than to continue his vain experiments to transmute base metal into gold. For the next six years Böttger worked under Tschirnhaus's supervision, and by the time the latter died in 1708, the two of them were producing both stonewares and porcelains in the laboratory at the Albrechtsburg Schloss at

Meissen, near Dresden. Two years later a factory had been established at Albrechtsburg which was to become famous as Meissen. The two scientists had discovered the secret, the arcanum, of hard-paste porcelain. They mixed together in the correct proportions the infusible clay, the decayed granite in its purest form known to the Chinese as kaolin, and the feldspathic rock which the Chinese called *petuntse*, meaning little white bricks.

In the years that followed, the factory, which Böttger founded and managed until his death at an early age in 1719, produced copies of both blue and white and polychromatic Chinese wares. These proved very popular, being greatly superior to the faience wares, confusingly known as *prozellan*, produced by the other German potters of the period. As a result, it provided the Elector of Saxony, the factory's chief patron, with considerable revenues.

The secret leaked out

The arcanum did not remain the sole property of Meissen for very long. It had to be entrusted to those who were responsible for preparing and mixing the ingredients of the body and the feldspathic glaze that fused into the matt biscuit body and transformed the whole into a sleek material. Then those who controlled the kilns had to master the important technique of firing the body and the glaze. Inevitably, sooner or later, someone who had learnt these secrets would come to realize that the arcanum was a saleable commodity. At some time prior to 1718, Böttger's kilnmaster, Samuel Stöltzel, left Meissen, turned up in Vienna and imparted the knowledge he had acquired to Claude du Paquier. This enabled du Paquier to open a porcelain factory in Vienna in 1719. By the middle of the century the arcanum had been disseminated throughout the Holy Roman Empire. Under the patronage of emperors, princes and dukes one factory after another was founded to produce porcelains. Frederick the Great founded a short-lived factory in Berlin in 1752 and Duke Charles I of Brunswick started the Fürstenburg factory, which is still producing today, in 1753. Prince Max Joseph of Baveria founded the still-famous Nymphenburg factory in the same year and the Duke of Würtemberg set up a factory at Ludwigsburg in 1758. The products of these and half a dozen other German factories were soon replacing silverwares on the tables of the wealthy at home and abroad, and eventually also replaced the pewterwares on the tables of the less well-to-do.

Soft-paste porcelain

The search for pottery that would more closely resemble the Chinese imports than the faience wares of the seventeenth century was not, of course, restricted to Germany. It had led to the production in France and England of what we call soft-paste porcelain. This is sometimes known as frit porcelain or, because of its great popularity at one time in France, as *porcelaine de France*. The French themselves referred to it as *pâte tendre*, and this was first produced, possibly as early as 1673, at Rouen. It would seem that some potter there concluded from studying the Chinese porcelains that they were a form of glass. So he mixed together the constituents of glass, fused them and ground down the resulting frit to a powder. This he then mixed with a white clay and fired the admixture in the kiln. The resulting pottery was very pretty and delicate, and remained popular in France over a long period. Soft-paste porcelain had, however, a number of inherent drawbacks. It was difficult to make, being liable to collapse in the kiln even though it was fired at much lower temperatures than hard-paste porcelain. It was also very fragile, and liable, as W. B. Honey describes it in his book *French Porcelain*, 'to crack at the touch of hot water'. It was also much softer than the true porcelains it sought to imitate, and can be scratched with a file, a test which one would not, however, recommend for identifying valuable antiques. If a broken piece of soft-paste porcelain is inspected with a glass its mealy texture can be seen, quite distinct from the sleek texture of true porcelain.

It is rather surprising that the French potters continued to produce soft-paste porcelain until the last quarter of the eighteenth century, despite the fact that a paper was published there in 1736 describing in considerable detail the methods which the Chinese used to make their hard-paste porcelains. This paper was based on a letter written in 1712 by a Jesuit missionary to

China, Père d'Entrecolles, who, during his visits to Ching-te-chen 'to administer to the spiritual needs' of his converts, interested himself 'in the manufacture of this beautiful porcelain, which is so highly prized . . . '. There followed a very detailed account of the way in which the Chinese potters worked and the materials they used.

The addition of bone

Not only did such famous French factories as Saint-Cloud and Sèvres continue to make soft-paste porcelain, but so, too, did most of the English factories such as Bow and Chelsea. Indeed, in Britain we have never adopted hard-paste porcelain, and today only one English factory, Royal Worcester, produces it. This was probably because one of the main drawbacks to soft-paste porcelain, its liability to collapse in the kiln, was overcome when in about 1744 the Bow factory began to add calcined bone ash to a white kaolinite clay. It is difficult to imagine how anyone came to consider using this unlikely ingredient, and one can only imagine that it must have been the outcome of a long series of empirical experiments during which every imaginable substance was tried. The addition of bone ash proved, however, to be most advantageous. It not only provided the much-desired stability during firing but it also produced a harder, stronger body, and after the glaze had been fired the resulting bone china proved to be markedly more translucent even than hard-paste porcelain. Soon all the leading factories in England followed the lead of Bow, and the tradition of English fine bone china was established.

The adoption of soft-paste porcelain in France and England had a marked influence on the style of pottery these two countries produced. To provide adequate strength, wares made from soft-paste porcelain had to be more substantial than the German hard-paste porcelains. As a result, the French and English potters relied more, as W. B. Honey points out in his book *English Pottery and Porcelain*, 'upon lively and fanciful painting rather than on the plastic extravagances of the Germans'. We see this exemplified in the products of the Chelsea and Bow factories from the 1740s onwards. And even today much English china is decorated with lively and fanciful 'painting'.

Early days at Worcester

The Worcester factory approached the problem in a different way, evolving a new porcelain formula using Cornish soapstone, which was very similar to the *petuntse* which Père d'Entrecolles described as one of the main ingredients used by the Chinese potters. The men who were probably responsible for this innovation at the Worcester factory were Benjamin Lund, a manufacturer of brass and copper, who was granted a license to search for soapstone near the Lizard in Cornwall, and William Cookworthy, a young apothecary from Plymouth, who first started a pottery factory in

Fig. 4.5. An eighteenth-century porcelain mug from Royal Worcester, with under-glaze blue painting (by courtesy of the Dyson Perrins Museum, Worcester)

his home town and later transferred it to Bristol. His Bristol factory was closed down in 1752 and the firm was taken over by the Worcester Porcelain Company. It is known that he had read Père d'Entrecolles' account and spent some years searching in the West Country for materials similar to the *petuntse* and kaolin which the missionary had described. He subsequently took out a patent in March 1768 for the manufacture of porcelain. The section of this patent which refers to the raw materials he employed shows how close he had come to the Chinese arcanum. 'The materials', it states, 'of which the body of the said porcelain is composed are stone and earth or clay. The stone is known in the countries of Devon and Cornwall by the names of moorstone and growan, which stones are generally composed of grains of stone or gravel of a white or whitish colour, with a mixture of talcy shining particles . . . All these stones, exposed to a violent fire, melt without the addition of fluxes into a semi-transparent glass . . . The earth or clay, for the most part, lies in the valleys where the stone forms the hills. This earth is frequently very white, though sometimes of a yellowish or cream colour.' So large were the deposits of kaolinite in Cornwall which Cookworthy discovered that to this day they still provide not only the English potteries with a vital raw material, but potteries abroad, such as Royal Copenhagen.

Cookworthy's researches enabled the Worcester factory to produce much thinner wares than its competitors, and its underglaze-painted blue and white porcelain was very fashionable in the eighteenth century (Fig. 4.5). Today, of course, these wares are very desirable collectors' items.

Polychrome and pâte

In the eighteenth century, European potters not only produced bodies and glazes identical to those made by the Chinese or plausible imitations of these, but initially also derived their decoration from China, adopting both Chinese motifs and established techniques. The Chinese had demonstrated many ways in which pottery could be embellished. Besides under-glaze painting they had decorated their porcelains with polychromatic designs, applying a large pallet of enamels over the glaze and firing these at temperatures low enough not to mar the colours. They had carved their pottery, imitated carving by imparting decoration from a mould, applied cast decoration to their pots including the delicate reliefs which the French call *pâte-sur-pâte*, or paste-on-paste, and enriched their pieces with intricate pierced decoration. In copying the Chinese motifs, or more often adapting these to their needs, European potters had little or no appreciation of their mystic significance, and as a result they frequently mixed their metaphors. However, few, if any, of their customers would have been aware of this. English potters have, since they began to produce bone china and porcelain, continued to draw inspiration from Chinese potters. Today, some of the more popular patterns of tablewares produced by the English factories still show a marked Chinese influence, while others are unashamed chinoiseries.

Wedgwood and jasper ware

One English potter, however, towards the end of the eighteenth century produced a type of pottery the appearance of which owed nothing to the Chinese. This man was Josiah Wedgwood, and what he eventually put on the market, after endless experiments, was a type of unglazed stoneware, an intrinsically coloured material, which he called jasper.

To understand what prompted Wedgwood to make his jasper wares one has to appreciate what an impact the discovery of the long-buried Roman cities of Herculaneum and Pompeii had on the arts and crafts of eighteenth-century England. By those who patronized the arts, the early part of the eighteenth century was conceived of as another Augustan age, a reincarnation of an idealized classical era. With the discovery first of Herculaneum in 1709 and then of Pompeii under its overburden of volcanic ash in 1748 it became more and more apparent as excavations continued that the ideas about classical artefacts handed down by the artists and architects of the Renaissance were misconceived. Classical art and architecture, it seemed, were not baroque, but altogether cooler and more restrained. So English designers, the Adam brothers in the forefront of them, copied the motifs from the plaster decoration found on the buried Roman walls and decorated the facades of buildings, the interior walls and ceilings, carpets and furniture with a profusion of delicate arabesques and stylized honeysuckle. The silversmiths turned teapots into tall funerary urns, inappropriately adorned with a handle and a spout. The potters, on the other hand, seem to have found themselves at a loss to know what to do about this change in taste. After they had shrugged off the Chinese influence to some extent, they embraced a wide variety of styles – baroque, rococo, Louis Seize – with varying success, but somehow the style which we call Adam did not seem to accord with the sleek surfaces of bone china and porcelain, It was Wedgwood who found the answer, who realized that what was needed was a new body, a body with a plaster-like texture.

Josiah Wedgwood had been born into the potter's craft, being the twelfth child of Thomas and Mary Wedgwood, who together ran the Churchyard Pottery at Burslem in Staffordshire. When Josiah Wedgwood grew up he entered into partnerships with a number of Staffordshire potters and began to take an increasing interest in the technology of his craft. One of the innovations that resulted from this interest was a cream earthenware which earned him royal patronage and which, as a result, he called Queen's Ware (Fig. 4.6). In 1762 he had, as far as his business future was concerned, a happy accident. He injured his leg and was confined to bed. To relieve his boredom his doctor introduced him to Thomas Bentley, a textile merchant and a classical scholar with a great enthusiasm for the neo-classical style. Wedgwood was caught up in Bentley's enthusiasm. It was not, however, until seven years after their first meeting that the two men went into partnership, a partnership that was to result in the production of a pottery of an entirely new kind.

Wedgwood might have been content to cater for the tastes of his times by making his cream wares in neo-classical shapes, but he decided that to produce wares that had the right neo-classical flavour he needed to produce a new body. This idea did not immediately bear fruit. The Wedgwood archives reveal that some 4000 experiments were carried out between 1771 and 1775, when the first jasper wares were put on the market. Indeed, it was not until 1779 that all the problems had been solved. As early as 1768, however, it could be seen in which direction Wedgwood was

Fig. 4.6. An eighteenth-century Queen's Ware coffee pot from Wedgwood (by courtesy of the Wedgwood Group)

moving. In that year he put on the market his first 'antique vases, urns etc. of black porcelain or artificial basaltes'. This black basalt ware was a Wedgwood improvement on an established Staffordshire ware called Egyptian black, which had been produced from local clay that had been coloured by coal sediments from the seams adjoining the clay deposits. This material, which Wedgwood burnished to a deep bronze colour, proved appropriate for the production of neo-Greek pottery, which he decorated with red enamel. It had some commercial success, but he rightly appreciated that this sombre pottery was likely to have only a limited appeal. What he was looking for was something more colourful and a body that closely resembled in its surface texture the plaster decoration on Roman houses and public buildings or the white marble that the Roman sculptors employed.

Those 4000 experiments carried out at the Etruria factory which Wedgwood and Bentley had opened eventually produced a body with the desired texture, but there still remained the problem of producing the body in the range of colours Wedgwood had envisaged. The addition of the necessary metallic oxides affected the firing properties of the body, and to overcome the problems this presented took another four years.

The body which he eventually evolved he described as being 'as fine as Parian marble but infinitely more durable'. This deliberately under-fired stoneware containing some 50 per cent of barium sulphate was integrally coloured in various shades of blue and green as well as black and white.

The first use which Wedgwood made of his jasper was to produce imitations of the agate and shell cameos so popular at the time as jewellery. These were designed for him by a number of talented artists, the most famous of whom was John Flaxman. To reproduce these cameos in quantity, open moulds were made into which slip was poured. When the slip had dried in the right consistency the cast was removed from the mould and the resulting low relief, usually of white jasper, was applied to a slab of coloured jasper with water and the cameo was then fired in the usual way. These imitation cameos have again been produced by the Wedgwood factory in recent years and have again proved very popular.

Soon, however, Wedgwood was beginning to produce more ambitious pieces, vases and candlesticks, jugs and bowls, the most famous of which was the jasper imitation of the Portland Vase, copied from the cameo glass vase brought back from Italy to England in 1783. These wares were decorated with relief motifs similar to the cameos, though more elaborate, and also cast in pitcher moulds. These white jasper reliefs depicted neo-classical figures, the acanthus motifs of classical architecture, fluting,

80 The discovery and development of the potter's craft

Fig. 4.7. A jasper 'Homer' vase from Wedgwood, white on dark blue, modelled by John Flaxman in 1784 (by courtesy of the Wedgwood Group)

lobbed borders, swags and vine leaves and, of course, the anthemion, the stylized honeysuckle so beloved of the Adam brothers (Fig. 4.7).

During the eighteenth century, pottery had become relatively inexpensive and had taken its place on the English dinner table.

Also, as more people began to take tea and to adopt the habit of snuff-taking, the potters catered for these new foibles, and the craft thrived.

The nineteenth century

The nineteenth century saw the growth of an opulent middle class whose wealth was generated by the Industrial Revolution. As a result, the demand for fine pottery increased dramatically, and what had been a craft became an industry. Large factories were built and the workforces of these were numbered in hundreds. It was thought largely a matter of a change of scale. Whereas in other industries machines increasingly replaced manual operations, the potteries remained, as indeed they still do to this day, very labour-intensive. Such mechanical aids as plate- and cup-making machines have only been adopted on a limited scale in quite recent times. In factories producing high-quality wares the making is still done by hand, and the decoration laboriously applied, brush stroke by brush stroke.

Besides the demand for a greater volume of production, the nineteenth century also brought a demand for novelty. It was an innovative century, and the nineteenth-century customer had an insatiable appetite for variety. New bodies were introduced, new methods of decoration were sought, new designs appeared in their thousands, and, if many of them lacked the elegance of the best wares produced in the previous century, these sometimes mawkish creations appealed to the taste of the times.

It would be impossible in a book intended to give an overall view of pottery and glass even to attempt to detail all the innovations of the nineteenth century. However, a brief description of some of them will suffice, to indicate just how resourceful the potters working between 1800 and 1900 were. Anyone who wishes to make a more detailed study of their achievements will find a large number of specialist books on the subject. They can also visit the major potteries, most of which have very comprehensive museums. Most local and national museums also have collections of pottery from this fertile period in their showcases.

Perhaps the two most interesting new bodies produced during the last century were ironstone and parian. The description 'Mason's Ironstone' is still to be seen on pottery made today by this firm, which was started in the eighteenth century by the Yorkshireman, Miles Mason, who, after working as a glass merchant and a chinaman in London, opened a factory at Newcastle-under-Lyme. It was his son, Charles James Mason, who registered the patent for ironstone in 1813. This consisted of 'a slag of ironstone, pounded and ground in water in certain proportions with flint, Cornwall stone and clay and blue oxide of cobalt'. This new body was usually decorated with Japanalia, presumably an imitation of the popular wares which Miles Mason imported from the Far East. The original firm failed, but after a series of takeovers the name was acquired by Wedgwood in 1973 and they have proceeded to put the Mason name back on the map.

Parian ware took its name from the name of the marble found on the island of Paros, which was used by the sculptors of classical Greece. Like jasper, it was created to imitate the materials used by the Greeks and Romans, and was inspired by the neo-classical movement that spilled over from the eighteenth century into the nineteenth. Parian ware was an unglazed biscuit porcelain which the Copeland factory introduced some time between 1842 and 1846. It is recorded that 'the first idea of imitating marble in ceramic manufacture originated with Mr Thomas Battan, the artist directing the extensive porcelain manufactury of Mr Alderman Copeland at Stoke-on-Trent . . .' Parian was used by the factory mainly to reproduce sculpture in the neo-classical style.

Some of the most widely used decorative techniques employed by nineteenth-century pottery manufacturers had been evolved earlier, but they were only fully exploited after 1800. The most popular decorative technique was transfer printing. Tens of thousands of pieces decorated with the famous Willow Pattern (Fig. 4.8), produced by half a dozen factories, must have come onto the market, the chinoiserie motifs printed from a copper plate in cobalt blue ink on to a transfer that was then applied usually over the glaze, but more rarely under it. The huge success of this pattern was, no doubt, due not only to the charming representation of a Chinese scene, but also to the legend it illustrated of a mandarin's daughter, promised in marriage to an

The nineteenth century 83

Fig. 4.8. A modern version of the traditional willow pattern (by courtesy of the Wedgwood Group)

old merchant, who runs away with her young lover. The couple eventually escape from the wrathful father by being transformed into the two doves incorporated into the design. Transfer printing was also extensively used to produce the outlines of patterns which were subsequently filled in by hand-painting in enamels. The use of polychromatic painting over the glaze became increasingly popular as the nineteenth century progressed, and naturalistic designs depicting landscapes, floral tributes, romantic scenes of home life and country folk going about their labours replaced the imitations of Eastern motifs favoured by the under-glaze decorators of the eighteenth century (Fig. 4.9). There was also a revival of the brashly coloured tin-glazed majolica style during the 1850s, which started at the Minton factory and was copied by a number of Minton's competitors.

Fig. 4.9. A Royal Worcester cabinet plate, a fine example of polychromatic on glaze decoration from the nineteenth century (by courtesy of the Dyson Perrins Museum, Worcester)

Minton was indeed one of the great innovators. They brought over artists trained at the Sèvre factory in France to enable them to produce replicas of the Sèvre designs, and it was also from this factory that they borrowed the technique first used by the Chinese potters, which is known as *pâte-sur-pâte*. This technique consisted of building up slip, layer by layer, on the surface of a vessel and then carving the slip to produce partially translucent designs of great delicacy. The most beautiful of these classical figures in flowing drapes were created by Louis Marc Solon, examples of whose work can be seen in the museum at the Minton factory.

Minton were also responsible for introducing a variation of *pâte-sur-pâte* known as pastc gold decoration and the engraved gilded style of decoration known as acid gold.

The twentieth century

The twentieth century has been one of consolidation rather than of innovation. Control of raw materials has become more exacting, the knowledge of the chemistry of bodies and glazes has increased and kiln technology has become more sophisticated. Some mechanization has been introduced in the making department and the glazing shop, and, of course, new generations of designers have introduced new shapes and new decorative motifs. Transfer printing has been replaced, except for one-off pieces, by silk screen printed or lithographic transfers. Alongside the factory production there is today a great deal of studio pottery being produced by craft potters, most of whom look to the pottery of the past for their inspiration, particularly to the beautiful pottery produced in China, Japan and Korea. Bernard Leach, who learnt the craft in Japan, was probably mainly responsible for this revival. Today most towns have a studio potter, and, though their work varies considerably both in quality and inspiration, they are helping to keep interest alive in a craft that goes back to prehistoric times.

5 Making pottery today

Clays, bodies and glazes

Pottery is made from decomposed granite, which has become transformed into what we know as clays. These clays contain varying quantities of the white mineral kaolinite, consisting of aluminium oxide, silica dioxide and water. Silica dioxide is, of course, the chemical name for quartz and this, which is one of the main constituents of granite, is the mineral which is fused to make glass. It can be appreciated, therefore, that pottery and glass are, like second cousins, loosely related.

Clay is a plastic material, which means that it can be readily shaped or moulded. This plasticity results from the fact that the kaolinite particles of which clay is composed are plate-like, and these platelets slide over each other, lubricated by the water present in the material.

Cornish clay

Though kaolinite is present in all clays, the quantity of it varies considerably. The nearer a clay deposit is to the granite formations from which it is derived, the purer it is likely to be, that is, the more kaolinite it is likely to contain and the fewer impurities will have been absorbed by the clay to contaminate it. Bodmin Moor in Cornwall is the source of the purest clay in Britain, and one sees the streams in the area running white, the water coloured by kaolinite particles from the adjoining beds of pure clay. The clay in the area round Stoke-on-Trent, which we call The Potteries, is, on the other hand, impure. Much of it is red in colour due to the presence of iron, but some of it, it will be recalled, is black due to the presence of coal from the nearby seams, and it was this black clay which the local potters used to make their Egyptian Black wares and which Wedgwood used to make his Black Basalt.

It was the existence of very large deposits of pure clay in many parts of China which led the Chinese to produce at a very early date the translucent white pottery which we call porcelain, one of the main constituents of which is the white clay they called kaolin. Today the majority of tablewares are made from white clays with a high kaolinite content, but the studio potters find the less pure clays more interesting to work with. These impure clays often have better working qualities on the wheel, and result in wares which are less sleek than the factory-produced pottery and more to the taste of the craftsman-potter and his customers.

What is pottery?

It would perhaps be useful at this point to define what we mean by 'pottery', for the word is often misunderstood, its meaning limited to low-fired earthenwares. It is, in fact, one of two words which are used to describe all those wares which are produced by driving out the water from clay by heat. The other word used to describe all these diverse products of the potter's kiln is 'ceramics'. These two generic words cover the whole spectrum, from the most delicate of bone china to sanitary wares and plant pots. They include the porcelains from Meissen and the Royal Copenhagen factory as well as inexpensive stoneware tea and coffee sets, the exquisite wares which the Chinese produced during the Sung dynasty and the work of such fine studio potters as Bernard Leach.

As has been stated, the clay plays a vital role in determining the type of pottery that is produced from it. The red clays found in The Potteries lent themselves to the production of earthenwares and stonewares. The clays of Devon and Dorset proved suitable for the production of jasper wares, and so on. But it is not only the clay which a potter uses that dictates the nature of the body from which he fashions his plates and cups, bowls and vases. What he adds to the clay and the temperature at which he fires his body also determines the type of pottery which he produces. Though there are many sub-types of pottery, all the wares which can be produced by dehydrating clay can be divided under two main headings. There are, broadly speaking, earthenwares and stonewares, and the essential difference between these results from the temperature at which they are fired.

Earthenware

If a potter puts clay into a furnace at a temperature of between 800 and 1000 degrees Centigrade this is high enough to drive out all the water present, not only free water but also the water chemically bound into the kaolinite. This will produce a pot, which can be fairly easily broken into virtually indestructible fragments, a fact which the millions of potsherds found on archaeological sites bear witness. This material, called by the descriptive name earthenware, is not only relatively fragile but is also porous. Earthenwares are usually covered with a glassy skin, known as a glaze, to render them waterproof. An unglazed form of earthenware is called terracotta, which simply means cooked earth.

Incidentally, some earthenwares are only fired once. They are not biscuit-fired, but the glaze is applied to the body after it has been formed and dried. The wares then go into the glost kiln and the body and glaze are fired at 1100 degrees Centigrade in one operation. In the past, this technique has been associated with rather cheap wares, the glaze on which often exhibited porosity. Recently, however, Poole have introduced a range of vases made in this way which are very attractive and do not exhibit surface blemishes. The only drawback to this technique is that the colour range is limited. Strong-coloured glazes, particularly reds, will not stand up to the high-temperature firing in the glost kiln.

Stoneware

To produce stoneware, clay is placed in a kiln at a temperature of 1250 degrees Centigrade or higher, with the aim of forcing the clay to fuse to form a vitreous mass. Even this temperature would not be high enough to fuse pure kaolinite, but to induce this the potter adds powdered feldspathic stone to his clay, and this acts as a flux. Stonewares are impervious to water, and are much harder and much stronger than earthenwares. They probably got their name because not only do they have, in their unglazed form, a texture like stone, but they also ring like stone it they are tapped. Only relatively few stonewares, jasper for example, and some studio-produced wares, are left unglazed. When a glaze is added to a stoneware this is done to improve its appearance and to make it easier to clean, not in order to waterproof it.

As has been said, these two main types of pottery can be subdivided. Earthenwares which have a salt glaze are variously called faience, Delftware and maiolica, while slipware is earthenware decorated with slip. The main types of stoneware are what are commercially described as stoneware, jasper, porcelain and bone china. While earthenwares are mainly differentiated on the basis of their decoration, the different types of stoneware result from different ingredients in their bodies producing quite distinct physical characteristics, and to their being fused at different temperatures.

In a modern specialist pottery shop the majority of stock will be what are commercially known as stonewares of one kind or another, including Wedgwood's jasper ware, porcelain and bone china.

Stoneware had been described as a ware which is part-way between an earthenware and a porcelain. It consists of clay mixed with a fusible stone which has been vitrified at a high temperature in the kiln to render it impervious to water. It lacks, however, the translucency which is associated with porcelain. When we come to discuss glazes it will also be seen that the relationship between the body and the glaze of a stoneware is different from that between the body and glaze of a porcelain.

Porcelain

The Chinese made their porcelain from two materials – clay with a high kaolinite content and white China stone, which they called *petuntse*, consisting largely of feldspar, a silicate of aluminium in which soda and potash are also present. China stone is fusible whereas the kaolinite clay is infusable, and what happens in the kiln is that the China stone melts to form a cement that binds together the infusible particles of the clay.

It may be recalled that the soft-paste porcelains, first created in France at the end of the seventeenth century to imitate the imported Chinese wares and widely used in England as well as France throughout much of the eighteenth century, consisted of a powdered glass frit combined with clay. Because the glass melted rapidly in the kiln into a liquid state there was a tendency for these wares to collapse in the furnace. The German potters, however, from the early eighteenth century onwards used the Chinese formula to produce the hard-paste porcelains, which are much more stable in the furnace and considerably stronger. At some stage Continental potters began to add quartz to the two other ingredients. Royal Worcester, which is the only pottery in the UK to produce hard-paste porcelains today, make their body from 50 per cent china clay, 25 per cent feldspathic stone and 25 per cent quartz, and this is also the formula which Royal Copenhagen have adopted. The quartz is added to make the melt more viscose, that is, less fluid, thus preventing the body becoming too soft during firing, resulting in distortion. What happens in the furnace is that the quartz fuses somewhat sluggishly over a wide temperature range as the pots that are being fired travel through the furnace. Besides white porcelains, some Continental factories produce a black porcelain known as *porcelaine noire*. The colour results from the inclusion of metallic oxides in the clay and from firing in specially designed kilns.

Bone China

Bone china was an English innovation, probably introduced at the Bow factory, though some authorities attribute it to Chelsea and others to Spode, and dating from the middle years of the eighteenth century. Today all the leading English factories making fine china produce bone china, the body consisting of 25 per cent china clay, 25 per cent Cornish stone and 50 per cent calcined bone ash made from cattle bones imported from South America (Fig. 5.1). Cornish stone is, to all intents and purposes, the same as the *petuntse* used by the Chinese potters. The role of the bone ash is still not precisely understood, but it has a number of functions.

Fig. 5.1. The raw materials for English bone china in the slip house (by courtesy of Royal Doulton Tableware Ltd)

Not only does it make the behaviour of the body during firing much more predictable, it also increases the mechanical strength of the fired body, and makes it pure white in colour.

It might be assumed that hard-paste porcelain, because it is fired at very high temperatures, would be stronger than bone china, but in fact bone china is the mechanically stronger of the two. It is, however, possible to increase the mechanical strength of porcelain by varying the ingredients, and Royal Worcester claim to have achieved this. Hard-paste porcelains can also be made that are highly resistant to heat shock, and these can be used to make oven-to-table wares. These are made from a material called cordalite, which contains aluminium oxide. Bone china is not suitable for this purpose. Hard-paste porcelains have very hard glazes fired at around 1400 degrees Centigrade, and these glazes are very resistant to scratching. On the other hand, porcelain is neither so white nor so translucent as bone china.

Glazing

The subject of glazes is a very complicated one. Whole books have been written about its chemistry, but these are, of course, of interest only to potters. For the purpose of this book an attempt will be made to explain the subject briefly in simple terms.

The term 'glaze' derives from the word 'glass', and the act of glazing consists of covering a pottery body with a glass-like outer skin. While the glazes added to earthenwares are both decorative and functional, rendering them impervious to water, the glazes applied to stonewares are almost entirely decorative. An unglazed jasper jug functions just as well as a stoneware jug that has been glazed. In the case of ordinary stonewares and earthenwares a coloured glaze is usually used, and this is very often the only decoration applied. On porcelains and bone china wares it is usual to apply a white glaze, and this contributes to their translucency.

As will be recalled, the potters of Delft in the latter half of the seventeenth century applied a tin glaze to earthenwares to produce a plausible imitation of the porcelains which the Dutch East India Company imported from China. Delftwares were imitated by the German potters of the period and were also produced in England, initially by emigré Dutch potters. They are variously known as English Delftwares or maiolica, while in other countries they are often called faience. The inclusion of tin oxide in a mixture of sand, lead oxide and sometimes small quantities of potash produced an opaque white glaze which covered the wares somewhat as a glove covers a hand. The glazes were an outer coating, unlike, as we shall see, the glazes which the Chinese applied to their porcelain bodies. Coloured tin glazes were also used on earthenwares; indeed, the earliest known glazes, those used by the Assyrians to produce their brick mosaics, were coloured tin glazes. Other types were used on earthenwares, including the colourful lead glazes and the salt glazes usually restricted to storage jars and other kitchen pottery.

The glazes used for porcelains and for bone china wares do not just lie over the body, adhered to it but separate from it as do those on earthenware and stoneware. They combine with the body, and fuse into it and not onto it. This is because the glazes used for these wares are chemically very similar to the bodies. The reason bone china is so highly translucent is also due to the close relationship between body and glaze. The body consists of two types of crystal in its fused form, and these have well-matched refractive indices which are similar to the refractive index of the glaze. So rays of light passing through the material are not dispersed, but bent at a consistent angle, as they are by water or glass.

Preparing the ingredients

The first step in the process of making pottery, whether one is talking about tablewares, decorative wares or figurines, or whether these are going to be produced in bone china, porcelain, stoneware or earthenware, is obviously to prepare and mix the ingredients. In the case of bone china these consist of calcined cattle bones, mostly obtained from the Argentine by way of the Swedish glue factories, china clay brought from Cornwall or from France, and feldspathic stone, also from Cornwall. The bones, which amount to 50 per cent of the constituents of bone china, provide the refractory element which resists the action of the heat in the furnace. They also, as has been pointed out, produce a very white body with a high degree of translucency. Bone has been a major constituent of china for over two hundred years, and that we still do not fully understand its function is not that surprising – after all, as one pottery manufacturer put it, nobody has ever been inside a furnace when pottery is being fired to see just what does happen. Even in this scientific age many of the practices in this industry, which were arrived at empirically, have become time-honoured and accepted without question. The 25 per cent china clay in bone china gives the resulting body plasticity, making it easier for the potter to shape it to his needs. It also provides stability in the furnace, forming, as it were, the skeleton of a ware during the firing. The feldspathic stone, which makes up the other

25 per cent of the constituents, serves as a flux. This means that it helps the clay body to flow when it is heated and eventually to vitrefy.

Once the constituents have been prepared, that is, the bone and the feldspathic stone have been ground to a fine powder, they are placed in a slip tank, where the powdered material is thoroughly mixed with clay and water to form a paste. Needless to say, the makers of fine china carefully check all these raw materials in their laboratory to ensure their purity. After mixing, the paste is passed through a fine sieve to ensure that any large particles are removed that would otherwise mar the wares and result in their rejection. The paste is then passed over a series of electromagnets that remove any iron that is present. It is important to remove every trace of iron because even in very small quantities it could colour the body.

Removing water

Next, the paste goes into a filter press where excess water is squeezed out of it to produce square or disc-like slabs of the

Fig. 5.2. A presser at work in Royal Doulton's Baddeley Green factory (by courtesy of Royal Doulton Tableware Ltd)

material (Fig. 5.2). These are then left for a period to cure. There is some dispute about what purpose this curing serves, and all potters do not agree about how long it ought to take. It is known that the Chinese potters left their body to cure for a long period, believing that this increased the plasticity of the material. At Meissen, too, it was traditional at one time to follow the Chinese practice and leave the porcelain body to cure for a period of eighteen months. At English potteries the body is left in this state for much shorter periods, and at Royal Copenhagen they say that the only reason for storing a body before using it is to give laboratory technicians the opportunity to check each batch to make quite sure that no alien substances are contained in it, and that the proportions of the ingredients are precisely what they should be. To fail to carry out such tests in a factory producing large numbers of wares every day to extremely high standards could lead to an unacceptable level of rejects further down the line, which could be economically disastrous.

Pugging

After storing and checking the body, whether it is a bone china or a porcelain body, it is passed through a 'pugging machine'. This simultaneously further refines the body and removes any air from it which could result in bubbles being incorporated in the pots that will be made from it. Figure 5.3 shows the body coming out of the pugging machine in the form of a continuous cylinder, which is cut with a wire into suitable lengths as it emerges. To make flatwares such as plates and dishes, and in some factories to produce cups as well, this body is used in the form in which it comes out of the pugging machine. The craft potter also uses clay of this dough-like consistency to throw pots on a wheel, or to build them up by the coiling technique. In a modern factory, however, many hollowares are cast in moulds, and to make these the clay must be in a state in which it will pour readily. To create this smooth, creamy casting material, known as 'slip', water is added to the body together with a small quantity of soda.

Fig. 5.3. The body emerging from a pugging machine (by courtesy of Royal Copenhagen)

Fig. 5.4. Semi-automatic plate-making at the Minton factory

Making plates

In a modern factory plates are made either on a semi-automatic or on a fully automatic machine. On the semi-automatic machine, known as a jigger, the potter sits at a wheel on which a mould has been mounted which will dictate the form of the top of the plate. He throws down onto this mould an appropriately sized lump of clay, and as this revolves and spreads out over the mould he brings down a metal arm, a profiling tool, which shapes the upper side of the clay into the form of the bottom of the plate. He is throwing the plate in the same way that a craft potter throws a dish, but he is doing it upside-down, and the mould and the profiling tool ensure that the plate conforms to the required pattern. This process is illustrated in Fig. 5.4, while in Fig. 5.5 a fully automated plate-making machine is shown. On this machine, clay is loaded onto a mould on a turntable, and from then onwards the process is carried out without human intervention. The clay is automatically spread over the mould by a revolving disc, and then the bottom of the plate is formed with a profiling tool. Cups can also be produced on similar automated machines. For long runs of the

Fig. 5.5. Fully-automatic plate-making at the Royal Copenhagen factory

same pattern, such machines are obviously labour-saving and the wares they produce are in no way inferior to those produced individually by a skilled potter.

Making hollowares

The production of hollowares by casting in a mould begins with the making of a model, which has to be over-sized because the body shrinks in the kiln. For example, before the model for a new teapot is created the firm's design studio, or a freelance designer retained by the factory, will have had to design a whole range of wares, for a teapot will be only one item in a set which will include plates and saucers, teacups, a sugar basin and a milk jug at least. Most designers begin by making drawings of these pieces,

Fig. 5.6. Blue Fluted half-lace porcelain tableware (by courtesy of Royal Copenhagen)

delineating both the shapes and the decoration which will be applied to them. Royal Copenhagen have in their archives the delicate and detailed water-colour drawings which Arnold Krog did as the first stage in introducing additions to the firm's original Immortelle service, which is known in England as Blue Fluted (Fig. 5.6), while shown here (Fig. 5.7) is one of the designers in the Royal Doulton studio making the drawings for a new pattern.

Fig. 5.7. A new shape on a designer's drawing board at the Royal Doulton Design Studio (by courtesy of Royal Doulton Tableware Ltd)

When the design has been created, the models are carved from clay with sculptor's tools, and it is from these models that the moulds are produced. Most wares are assembled from a number of castings, so first the model is inspected and the decision taken as to how best to cast it, and then it is cut up into the desired component parts. The body will be cast separately, as will the handle and the spout, while a separate mould will be produced to cast the lid. The various pieces then have plaster of Paris poured round them, and when this has set the block is cut open and the model removed. A sprue hole is created, the mould is reassembled and the slip is poured in through the sprue. The moulds have only a limited life, and new moulds have to be continually produced. The slip is left in the mould for about twenty minutes, by which time a layer of it will have set round the walls of the mould. The residual slip is then poured out.

Assembly

Next, the moulds are broken open, and it remains for the cast components of the teapot to be assembled. When these components have dried to a leather hardness they are, as it were, glued together with slip. Then the assembled pot is left to dry.

Sponge, fettle and tow

When the wares from the plate department and the casting shop have dried out, a series of operations are carried out on them known as 'sponge', 'fettle' and 'tow'. Women using damp sponges and small scalpels, called fettling tools, remove any bumps and lumps which mar the smooth surface of both flatwares and hollowares and smooth out the junctions between the assembled components of the hollowares. A tow, a mop usually of flax fibre, is used to smooth flatwares and to round off any sharp edges. After these operations all the pieces are individually inspected. As every operation involves cost, any sub-standard ware is either sent back for rectification or scrapped before any further work is carried out on it.

Biscuit-firing

Bone china is fired for twenty-four hours at a temperature of 1250 degrees Centigrade in the biscuit furnace. During the firing the wares are supported on 'bats' or 'saggers' made from a heat-resistant material (Fig. 5.8). Hard-paste porcelain is fired for a similar period at 900 degrees Centigrade. Bone china wares emerge from the kiln with a matt biscuity surface, and during the firing they will have shrunk by one-eighth their original size. The matt surfaces of the wares are then brushed to remove any coarse

Fig. 5.8. The completed pieces go into the kiln for firing

material from them and the wares are individually inspected once again. The wares are then loaded into a vibrating machine along with wood chips which smooth them to the desired degree (Fig. 5.9). Too high a polish is not desirable, because this would result in the glaze not adhering to the biscuit surface.

Applying the glaze

In British factories under-glaze decoration is very rarely used these days, so the next operation is to apply the glaze. At many factories, at Minton, for example, the glazes used for the flatwares are tinted blue and those for the hollowares are pink. The chemicals used to tint these glazes are subsequently driven off in the furnace and do not affect the colour of the fused glaze. The

102   Making pottery today

Fig. 5.9. Polishing the biscuit ware after firing (by courtesy of Royal Doulton Tableware Ltd)

glazes are differentiated in this way because they have a slightly different consistency. Those used for flatwares flow more readily and, of course, the consistency of these glazes is very critical, otherwise they will not adhere equally to all parts of the ware. The glaze may either be applied by dipping or spraying. For flatwares some factories now use ingenious automatic machines to apply the glaze. Such a machine is used at the Minton factory and, besides speeding up the process, this has the advantage that it avoids the possibility of fingerprints marring the ware when it emerges from the kiln. This is a very important consideration in this factory, which uses a lot of gold decoration. When a band of gold is applied to the rim of a plate the slightest indentation shows up. The machine in this factory sprays 0.225 millimetre (0.9 thousandths of an inch) of glaze on the top of the plates that pass through it. Then the plates are automatically turned over and the same thickness of glaze is then sprayed on the underside. When the flatwares come off this machine each one is inspected, and any areas that are not covered with glaze are touched up with an air-brush. Next the feet of the vessels are cleaned of glaze so that they will not stick to the pins or pegs that support them in the zircon frames in which they are placed before going into the glazing kiln. The wares stay in this kiln, still known as the glost kiln, for a period varying from five and a half hours to eight hours. The glaze on bone china is fired at 1100 degrees Centigrade and that on porcelain at 1400 degrees.

When the glazed pottery comes out of the glost kiln the marks left by the supporting pins or pegs are ground out of good-quality pottery, though cheaper pots often show these peg marks. A well-finished foot is a sign of a good-quality pot. The labour involved in grinding off the marks is only justified if the price of the finished ware is high enough. The blanks are now sleekly white, and will have that characteristic translucency of both bone china and porcelain. The surfaces of porcelains with their hard glazes can be mechanically polished. Rosenthal state that the rims of their cups are ground and polished six times.

Throwing and coiling

Pottery is still, of course, made by throwing it by hand on the wheel (Fig.5.10). Some factories, of which Wedgwood is one, employ craftsmen to produce wares in this way, but throwing is more often practised these days by individual studio potters working mainly in earthenware and stoneware. Studio potters also employ two other techniques of making pots, techniques which predate the invention of the potter's wheel. These are the coiling and slab methods. Coiling, probably the earliest method of making a pot, consists of first making a base from a disc of clay. Then a ball of clay is taken and this is rolled out on a flat surface. This snake of clay is then coiled above the base. A second coil is made and this is then coiled on top of the first. With wet hands the potter

Fig. 5.10. Throwing a vase on the wheel, the most skilful of all pottery crafts (by courtesy of the Wedgwood Group)

then smooths the coils inside and out so that they amalgamate, and then goes on piling coil upon coil, shaping the vessel with the hands until the pot is complete. (See Figs. 4.1(a) and 4.1(b).)

Slab pots are made from slabs of clay which have been rolled out to the required shape and size. These slabs are then assembled by the potter, again using water and hands to amalgamate the components into a whole. Slab pots tend to be square rather than circular in form.

6 Decorating pottery

Pottery is often already decorated during the course of manufacture. For example, some blanks will have patterns impressed on them by the mould. Plates may be fluted simply by the plate-maker placing a suitable mould on his wheel. Similarly, if flutes are carved on the model of, say, a teacup and a plaster of

Fig. 6.1. Making the cameo decoration for jasper ware at the Wedgwood factory. Body is pressed into a 'pitcher' mould made from fired pottery. Then, using a spatula, the intricately shaped ornament is eased out of the mould (by courtesy of the Wedgwood Group)

106 Decorating pottery

Paris mould made from this, the mould will impart the imprint of these flutes onto the slip that is subsequently poured into it. Even such delicate decoration as half-pearls in the form of a circlet round the rim of a plate or a saucer can be applied to the blanks from moulds. For some decorative wares the old technique of pressing clay into the mould by hand is employed (Fig. 6.1). A modern ware made by this technique is the elaborately pierced orange bowl produced by Wedgwood to a design in the firm's eighteenth-century catalogue of Queen's Ware.

In some instances further decoration may be added to a blank before the biscuit-firing. At the Royal Copenhagen factory one can see skilled craftswomen modelling tiny flowers by hand, which will be used to embellish some of the more elaborate pieces of the

Fig. 6.2. Hand-piercing the body before it goes into the biscuit kiln (by courtesy of Royal Copenhagen)

Flora Danica pattern. Another type of decoration which can be applied before a ware is fired is piercing. This is also a feature of patterns produced by Royal Copenhagen, not only Flora Danica but also the lace versions of Blue Fluted. The holes in the body, which produce a lace-like effect, are created by pushing a metal tool through the clay (Fig. 6.2) and then trimming the edges with a fettling tool. Needless to say, both these hand operations add considerably to the cost of a piece.

Another way of decorating pottery is to produce a combination of matt and glossy surfaces. This is a speciality of the Rosenthal pottery in West Germany. The effect can be achieved in two ways. Parts of the glaze can be etched with hydrofluoric acid, the rest being protected with a resist, or alternatively a matt glaze may first be applied, then those areas which are to be left matt are stopped off with wax and the ware glazed in the normal way.

Decoration and price

Whether or not a piece of bone china or porcelain has been decorated in the course of making, it will invariably have further decoration applied to it in the painting shop. This decoration may consist of no more than a gold or enamelled band round the rim, but many designs call for much more elaborate decoration. This can vary from polychromatic motifs applied from a lithographic sheet to the most elaborate paste gold work, and the price of tablewares, or, for that matter, decorative wares, is very largely dictated by the amount and type of decoration applied to them. A factory with a reputation to uphold will make only one quality of blank, the very best it is possible to make. So all the bone china blanks made at, say, the Minton factory or all the porcelain blanks that go into the painting shop at Royal Copenhagen will have cost the firm the same amount of money to make. But when those plates have been decorated the difference in price can be dramatic. A Minton plate decorated with a simple band of colour might retail at £10 ($15). The identical blank could, however, have been first decorated with acid gold and then with elaborate gold paste work, and the most decorative of plates in this firm's gold paste range was selling in 1982 for no less than £2225 ($3340) for a single plate. In addition, Royal Copenhagen make two designs of blue under-glazed tableware, the Blue Fluted, already referred to, and Blue Flowers. Both are painted by hand, but it takes a skilled decorator twice as long to paint a piece of Blue Fluted as it takes to paint the same piece in the Blue Flowers design. When one looks at the Royal Copenhagen catalogue one finds that this fact is reflected in the prices. Blue Fluted is almost exactly twice as expensive as Blue Flowers.

Hand-painted pottery

In the past all fine pottery was painted by hand. Metallic oxides were mixed with ground-up silica and the mixture was then fused

Fig. 6.3. Decorating a piece of Flora Danica over the glaze with enamels. In the background is a replica of the original wash-drawing from the eighteenth-century book (by courtesy of Royal Copenhagen)

to produce a frit, which was then ground to a fine powder and suspended in turpentine. This enamel was then painted on the pottery with the kind of brush that a water-colour artist uses, then the enamel was fired. One still sees a large number of painters decorating pottery at the leading factories both in the UK and abroad. At Royal Copenhagen, for instance, they employ no fewer than six hundred painters, and a similar number work at the Meissen factory. Some reproduce set patterns, and they become so skilled at doing this that they can repeatedly reproduce a design with great precision. Some special pieces are decorated freehand, the painter being given a degree of freedom in interpreting the pattern. The women artists who paint the Flora Danica pieces, though their interpretations must be correct from a botanical standpoint, nevertheless develop their individual styles (Fig.6.3). Some mechanical aid may be employed. The work may first be outlined by painting through a perforated sheet. Another help for the painter is a pricked pattern, which results in a series of dots being applied to the surface of the blank as a guide to the design. Even when such aids are employed, however, hand-painting still remains a time-consuming and therefore an expensive process.

Under-glaze painting

Though most bone china is painted over the glaze, some European porcelain is still painted under it. This has the advantage that the fused enamel is protected by the glass-hard glaze, but, as was mentioned in Chapter 4 the painter's palette is limited if he works under the glaze, as many enamels would be discoloured by being subjected to the high temperatures of the glost furnace. Only certain metallic oxides will withstand this heat, notably cobalt oxide, used to create the rich blue for which the Worcester Pottery, among others, was famous in the eighteenth century. This is the same blue which is used to decorate the Royal Copenhagen patterns, Blue Fluted and Blue Flowers. Other colours which can be used for under-glaze painting are green, produced by making a frit with copper oxide, and a rather rusty red shade, resulting from the use of an iron oxide frit. Towards the end of the eighteenth century, and increasingly during the nineteenth, the public favoured pottery decorated with a whole palette of colours which allowed floral designs to be rendered in a realistic way and popular paintings to be translated into decoration for pottery. And ever since then the English factories have concentrated on polychrome decoration over the glaze. The enamels can then be fired at around 850 degrees Centigrade, temperatures which they can happily withstand.

In-glaze decoration

A compromise between under-glaze and over-glaze decoration is a process known as 'in-glaze' decoration. Hand-painted or transfer decoration is applied to a glazed ware which then goes into the kiln at the temperature at which the glaze was fired, that is, 1400 degrees Centigrade for porcelain. The ware stays in the kiln for about twenty hours, during which time the glaze melts and the colour becomes absorbed in it. Only pigments that withstand the temperature of a glazing kiln can, of course, be used for in-glaze decoration. It is, however, possible to use a wider range of colours and even to in-glaze gold and platinum by employing a rapid-firing technique, that is, firing porcelain at 1200 degrees Centigrade for only an hour and a half. The advantage of in-glaze decoration is that, like under-glaze decoration, it is protected from wear.

Transfer printing

The technique of transfer printing which the Irish engraver John Brooks invented about 1750, when he was working for the short-lived Battersea Enamel Works, is still used today, though only for special pieces and short runs. The process consists of first producing a copperplate, engraved with the required design (Fig. 6.4). This plate is then 'inked' with a metallic oxide in suspension. A piece of tissue paper is then laid over the inked plate and plate and paper are cranked through a roller. The pressure applied to the roller imprints the design from the copper onto the tissue. This is, in fact, a simple printing process. The tissue paper is applied to the pottery, and after the ink has dried on the surface of the blank,

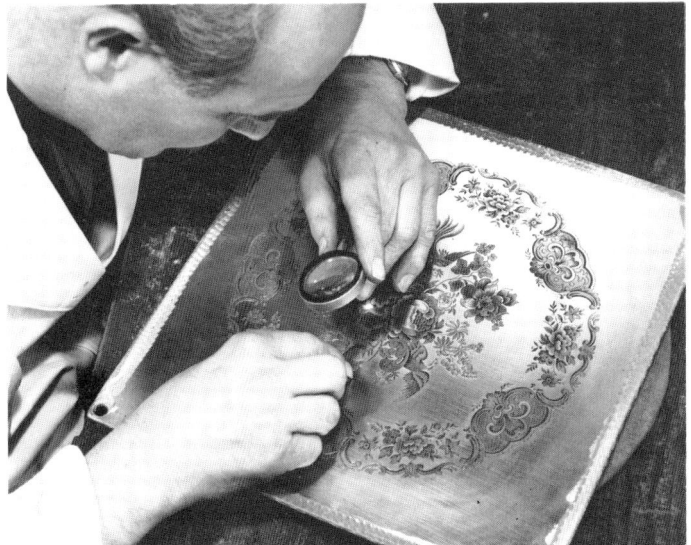

Fig. 6.4. Engraving a copper plate. This technique has now been largely superceded by lithographic printing

the tissue paper is carefully and laboriously sponged off. What is now left on the pot is a monochrome outline, which is then fired. Such well-known designs as the blue Willow Pattern were produced until quite recently by transfer printing. Later in the eighteenth century some polychromatic transfer-printed designs were produced, but these experiments were short-lived. Sometimes, too, the transfer print was subsequently elaborated by hand-painting, details being added in different coloured enamels.

Lithographic printing

Though similar in some ways to transfer printing, this recently introduced technique has a number of advantages over it. The most important of these is that it makes it possible to reproduce a polychromatic design in one operation with a minimum of labour. The lithographic sheet is also much easier to handle than the tissue paper used in transfer printing, and there is no need to spend time getting rid of the tissue by laborious sponging.

The production of the sheets used to transfer this decoration to the pottery is though, very expensive. These sheets are either produced by a lithographic process or by silk screen printing. If the silk screen technique, which tends to produce more muted shades, is used, many operations are required to obtain subtle colours. The silk screen process consists of transferring the colours to a piece of silk, individual transfers being used to stop off areas where particular colours are not required. Because of the cost of producing the sheets, lithographic printing is only applicable when large numbers of pieces are going to be decorated with the same design, so spreading the initial outlay. This is why transfer printing is still employed for short runs. Occasionally, too, it is still used to impart a period flavour. An example of this is the Mason's Ironstone pattern shown here (Fig. 6.5).

Lithographic prints are familar to model-makers, consisting of a sheet of plastic film on which the design has been printed. The designs are printed in enamels when the sheets are intended to be used for pottery decoration and these designs are protected by a backing sheet. All an operative has to do is to soak off this backing sheet and then apply the lithographic sheet to the cup or plate to be decorated. Even this simple operation does, however, call for some skill and experience, as the design has to be laid down in precisely the right position to ensure that the pieces in a service all match, and that, in the event that a coloured or a gold border is to be subsequently added, the spacing between the design and border is as it should be. If there is insufficient separation, it looks awful.

Once the transfer has been applied to the ware it is left to dry. Then the decorated ware goes into the kiln, which fires the enamel colours and burns off the plastic film. A piece of pottery decorated in this way resembles a piece that has been hand-painted, but with very little experience it is not difficult to recognize a printed design from a painted one, as there is something slightly mechanical about the look of it. A sure indication is, of course, price. If a piece that is decorated with an elaborate polychromatic pattern is modestly priced, the decoration must have been applied by printing. This is not to suggest that printing is not a very effective process, for this technique has enabled pottery factories to put many beautiful designs on the market at prices within reach of most people.

Fig. 6.5. Early Victorian design from Mason's Ironstone, one of the few modern patttterns decorated with copper-plate engraving (by courtesy of the Wedgwood Group)

Gold-enrichments

Gold-enrichment is frequently used on tablewares and decorative pieces of pottery. At its simplest, this consists of a band of gold round the rim of a plate, a saucer or a cup. This, whether it is a coloured band or a gold one, may be applied to the pieces with a brush, or if the pieces are produced in large numbers it may be applied automatically as the pieces revolve on a turntable (Fig. 6.6). Gold may also be used to enrich handles, lids, etc. and is put

Fig. 6.6. A flatware gilding machine. Colour banding or the application of gold bands is now automated in some factories (by courtesy of Royal Doulton Tableware Ltd)

on by hand with a brush. Gold in powder form is suspended in a suitable oil, and when the wares go into the kiln and are fired at between 850 and 900 degrees Centigrade, the oil is driven off and the gold fused onto the surface of the pottery. Either one or two coats of gold may be applied and fired, depending on the quality which a particular factory aims to achieve. The difference between a single coat and two coats is very noticeable, a single coat tending to have a rather brassy look, while two coats produce a deep, rich golden effect. On good-quality wares this is further enhanced by

Fig. 6.7. Hand-burnishing the gold with agate

hand-burnishing. This method of giving a metal a high polish is well known in the silver trade, and it consists of rubbing the metal with a piece of agate mounted on a stick (Fig. 6.7). On some decorative wares as many as four different-coloured golds may be applied to a single piece.

Acid gold

This is a technique invented in 1863 by James Leigh Hughes, and the patent rights were later acquired by the Minton factory. The technique remains very much a Minton speciality, but a number of other factories, Worcester and Rosenthal among them, also employ it, and it makes the gold band applied to pottery more interesting by creating a three-dimensional effect. The first step is to etch a design into the surface of the pottery. This is done by affixing a transfer print or a lithographic sheet of the required design to the ware (Fig. 6.8). This sheet consists of a black and white design, the black areas of which are printed with a compound that is resistant to acid. When this strip has been applied, the rest of the ware is covered with Brunswick black applied with a brush. Brunswick black is also resistant to acid. The ware is now immersed in a tank of hydrofluoric acid which bites into the glaze in those areas which are not protected by the acid resist. The ware is kept in the acid for a period long enough to allow the acid to eat into the glaze to a depth of 0.5 millimetre (2 thousandths of an inch). The acid is then washed off in cleansing

Fig. 6.8. Affixing a lithographic transfer to the ware (by courtesy of Royal Doulton Tableware Ltd)

tanks and the ware is dried. Then the Brunswick black and the lithographic resist is dissolved away in a solvent bath and the ware is again thoroughly washed and dried. The white surface of the ware now has a band of lace-like intaglio decoration running round it. Gold is now hand-painted over this decoration and fired. Then a second coat of gold is applied and this is fired in its turn and then burnished. The result is a gold-enrichment that also has a texture which imparts to it an additional richness. As may be imagined, this is an expensive process, not so much because of the cost of the gold but because of the amount of exacting hand-work involved.

Gold-paste decoration

This is the ultimate in rich decoration, and it is labour-intensive and time-consuming. Every piece to be decorated in this way is first very carefully inspected to ensure that it is perfect, as there would be no point in using a less than perfect blank. The technique is really a form of that style of decoration known as *pâte-sur-pâte*. Like so many pottery techniques, *pâte-sur-pâte* was first used by the Chinese potters, but the Sèvres factory in France adopted it in 1851 and it was brought over to the Minton factory later in the

Fig. 6.9. Minton's Shrewsbury Green pattern, decorated with elaborate paste gold (by courtesy of Royal Doulton Tableware Ltd)

century. Subsequently Minton developed their gold-paste form of decoration from it and it has remained the speciality of this factory ever since. At its simplest, it consists of a series of half-pearls of slip applied to a ware and then painted with gold. The pieces decorated in this way are very expensive; a 262-centimetre (10½-inch) dinner plate decorated with acid gold and simple gold paste cost about £65 ($100) in 1982. Some wares, however, are far more elaborate. Minton's famous H4230 pattern Shrewsbury Green (Fig. 6.9) is enriched with baroque motifs, scrolls and acanthus leaves, vases and friezes, all carried out in applied paste and enriched with gold. The 1982 price of one dinner plate in this pattern was £2225 ($3340).

The role of the inspector

One cannot emphasize too strongly the importance of the work of the inspectors, who check the wares at every stage in their production, in leading factories in the UK and on the Continent. It is clearly in the interest of these factories to ensure that expensive additional operations are not carried out on faulty blanks. So the inspectors who check these blanks after they come out of the biscuit kiln, and again when they emerge from the glost furnace, are making an important contribution to the economics of production. But this also works to the advantage of the retail trade and to the benefit of the ultimate customer. By reducing wasted effort, and by helping to reduce the rejection rate at the finished-product stage, inspectors make it possible for the factory to hold down its prices to levels that are more generally affordable than they might otherwise be.

When it comes to final inspection, though this may be primarily designed to maintain the factory's reputation in the marketplace, it is clearly in the best interests of the stockist that this should be as unrelenting as possible. Faulty wares, even if these are picked up by the retailer at the time of delivery, nonetheless involve him in considerable trouble and expense of obtaining replacements. If they are not spotted before they are sold they can obviously lead to customer complaints, something which no retailer likes to be involved in.

It is standard practice in every major factory producing bone china or porcelain wares for every piece to be inspected before it is despatched. The inspectors responsible will be looking for a variety of common shortcomings, such as black spots caused by foreign matter having got into the body or the glaze, unevenness or surface marks, discolouration, blemishes in the decoration, marks below the gold on rims, incorrectly placed printed decoration, etc. A ware that shows any of these defects will either be sent back to the factory to be rectified, where this is possible, designated as seconds or at worst scrapped altogether. Just how exacting the final inspection is can be judged by looking through a delivery of seconds from any factory with a high reputation for quality. On some of the pieces, it is true, the faults may be apparent, but on many the reason for their rejection will be almost imperceptible. Indeed, it is often the blue pencil mark put on the piece by the inspector that directs attention to some tiny blemish.

The fact that such rigid standards are set and normally adhered to does not, of course, mean that the retailer and his staff should take deliveries for granted. It should be standard practice for every consignment to be subjected to a further inspection on arrival. Inspectors, even the best of them, are only human and have their off-days, and, in the final analysis, it is the responsibility of the retailer to ensure that what he sells as perfect, is perfect.

7 Pottery figurines

Pottery figurines have a history going back to Neolithic times. Having discovered how to turn clay into pottery by firing it, early man shaped this new material into votive figurines, tributes to the mother-goddess and other deities. The pottery figurines, the pretty ladies produced by Royal Doulton and the under-glaze painted porcelain children from the Royal Copenhagen factory in Denmark or the Ladro factory in Spain, are, of course, a world apart from these primitive creations, though they still have pride of place in many houses as once the little goddess figures must have done in the crude huts of our Stone Age ancestors. The only pottery figurines made in the West that retain any religious significance, perpetuating that ancient tradition, are those rather crudely painted representations of the Madonna which one sees in the gift shops at Lourdes and other places of pilgrimage. The Chinese potters do, of course, continue to produce their traditional designs, many of which once had a votive significance, but tourists who buy them nowadays usually have very little, if any, idea of their symbolic role.

The English figurine, which is bought simply because it is attractive, dates from the seventeenth century. In the second half of that century salt-glazed figurines were produced by John Dwight of Fulham, who is believed to have employed the services of the famous wood-carver Grinling Gibbons to create his patterns. These white salt-glazed pieces continued to be made well into the eighteenth century, when, in the form of chimney ornaments, they were produced in considerable numbers in

Staffordshire. Notable among these ornaments were the famous pew-groups. The true ancestors of todays figurines were, however, those produced at the Meissen factory. The success of Tschirnhaus and Böttger in discovering how to produce porcelain provided an ideal medium for producing figurines, the mechanical strength of porcelain making it possible to produce delicate models with fine detail. Meissen, too, broke new ground by employing the services of leading artists of the day to create their designs, the best-known of whom was Johann Joachim Kändler, who trained as a sculptor and joined Meissen as a modeller in 1731.

In England the Bow and Chelsea factories began to produce figurines in soft-paste porcelain in the middle years of the eighteenth century. These were very much in the style of Meissen and the other Continental factories, such as Sèvres. There is a naive charm about these shepherdesses, milkmaids, dancers, men in stylish waistcoats and knee-breeches and women in fashionable furbelows, which make these early pieces so desirable to collectors. The introduction of bone china figures, which brought a reduction in firing problems and made it easier to produce more detailed and complex designs, unfortunately coincided with lower standards of modelling. There were, of course, exceptions. One thinks, for instance, of the bisque figures produced at Derby in the 1790s and Wedgwood's black basalts and jasper figurines, but, generally speaking, the figurines from the late eighteenth and early nineteenth centuries lack the charm and spirit of the earlier Bow and Chelsea pieces.

A century of contrasts

The nineteenth century saw the production of figurines in a host of styles and a wide range of materials, and there must have been at least one pottery figurine on every mantelpiece in Britain by the end of this period. Every pocket and every taste was catered for. Rather clumsy earthenware figures of fishwives and shepherdesses were produced alongside beautifully modelled Parian wares like the Venus from Worcester, now in the Dyson Perrins Museum, and the Prisoner of Love, made by the Belleek Pottery in County Fermanagh and now in the Ulster Museum in Belfast. Then there were the exotic English maiolica groups like the hen and cock modelled by John Henk for Minton, to be seen in the Minton Museum at Stoke-on-Trent, and, of course, there were some fine pieces cast in bone china at this time, such as the Triton modelled by M. Gautier for Minton.

Nostalgia sells

Two very popular products of the nineteenth century were the Staffordshire earthenware dogs and Toby jugs. Toby jugs were probably first produced by Ralph Wood in the second half of the eighteenth century, and are still produced in their thousands today. Indeeed, most of the popular pottery figurines (and Toby jugs really belong among these, as their utilitarian purpose – if any – was secondary to their decorative purpose) date back in style to the nineteenth century. Today's modellers, many of whom are extremely talented, have to produce pastiches to satisfy the public's taste for the nostalgic where figurines are concerned. Modern designs have proved over the years to have only limited appeal. If one looks through the book by Desmond Eyles and Richard Dennis on Royal Doulton figurines, one can hardly discover a single figurine that could be described as contemporary. The pretty ladies designed by Peggy Davies and Eric Griffiths in recent years wear crinolines, the nannie who sits sewing a quilt is a Victorian nannie, the huntsmen are timeless, as, too, are the Indian braves. We seldom get closer to modern times than the 1920s, to which period figurines such as Dawn and A la Mode belong. At Royal Copenhagen nearly 80 per cent of the designs they produce date back to 1900 or earlier.

The animal kingdom

Besides the human figure, one must include under the heading of pottery figurines a whole menagerie of animals and birds. Some of these are created by sculptors of great ability and rank as works of art. One would include among them the superbly modelled birds which Dorothy Doughty created for Royal Worcester. At the other end of the scale are the somewhat mawkish representations of pet dogs and horses which enjoy a very considerable popularity, and, of course, those wild duck plaques that at one time decorated the walls of so many homes. All these figurines attract collectors, and limited editions like Doughty's birds and Doulton's recently introduced Figureheads sell out in next to no time.

Moulding techniques

The technique of casting pottery figurines in a mould dates back for centuries. Early moulds were, however, what are known as single-valve moulds, half-shells into which the clay was pressed and which produced figurines with flat backs. Though such moulds were still in use in the eighteenth century, the double-valved mould, the two sides of which were joined to allow a figure in the round to be cast, dates back to at least the fifth century BC, at which time it was being used by the Chinese potters. This type of mould is used by producers of pottery figurines today, and also, as has been described in Chapter 6, for the production of hollowares. These plaster of Paris moulds have slip poured into them, the walls of the mould absorb the water, the slip sets evenly around the interior surface of the mould and excess slip is then poured off. The resulting figurine will, of course, show a seam, resulting from the inevitably slight mismarriage of the mould's two halves, but this is easily removed. Like hollowares, all except the simplest figurines are cast in parts, heads, arms, etc. being cast in separate moulds.

The designer

The quality of the final figurine depends, apart from the skill of the caster and the man or woman who paints it, on the artist who originally conceived it. These artists are often sculptors of some distinction, and, though they may make preliminary sketches, usually made the original model in clay. Figure 7.1 shows Eric Griffiths, Doulton's design director, modelling one of the figures for the firm's Image range in clay. He describes clay as 'a receptive

124 Pottery figurines

Fig. 7.1. Eric Griffiths, Royal Doulton's Design Director, making a clay model for a new figurine

material for a modeller to work in and naturally sympathetic to pottery'. When a model has been produced, oversize to allow for shrinkage in the kiln, the decision is made as to how it shall be broken down to produce the various moulds. The popular Boy with Calf, for instance, modelled for Royal Copenhagen by Christian Thomsen, is cast in nine parts, for each of which a separate plaster of Paris mould has to be produced (Figs 7.2 and 7.3).

Before it puts a new figurine on the market a large factory will produce a short run, perhaps decorating the figurine in a number of different ways. They will then carry out market research, showing the new design both to retailers and to groups of potential customers before going into full production. Even so, not all designs are equally successful. The public's taste in figurines is

Fig. 7.2. The nine moulds used to cast the parts of a Boy with Calf figurine (by courtesy of Royal Copenhagen)

fickle, and the endless variety of different styles of figurine on the market result from the efforts of the manufacturers to appeal to every taste and every pocket.

Materials and methods

Figurines are made in a variety of materials. Animals are often cast in stoneware or even earthenware. Royal Copenhagen and most of the Continental factories use mainly porcelain, and the UK factories bone china, though Royal Doulton also use a feldspathic pottery which they describe as 'fine china', which shrinks by 17½ per cent in the kiln compared with a 12½ per cent

Fig. 7.3. The different components of a group, like the Boy with Calf, are bonded together with slip (by courtesy of Royal Copenhagen)

shrinkage for bone china, a fact which the modeller has, of course, to be aware of. When the parts of a model have been assembled with slip (Fig. 7.4) they are allowed to dry. Some figurines also have additional decoration added, tiny hand-made flowers, for instance (Fig. 7.5) which are attached with slip. After the figurines have been left to dry they are sponged and fettled (Fig. 7.6) to remove seams and any other surface blemishes. Each figurine is individually treated, every one is carefully inspected and any exhibiting faults that can be rectified are sent back for additional attention. These figurines then go into the biscuit kiln and receive the firing appropriate to the type of body they have been cast in. After firing, the figurines are polished, and those that are to be painted under the glaze go to the painting department, where they

Fig. 7.4. Pouring the slip into a mould to produce part of a figurine at the Royal Doulton factory (by courtesy of Royal Doulton)

128   Pottery figurines

Fig. 7.5. Some figurines have hand-modelled flowers applied to them after casting. These are then painted in natural colours (by courtesy of Royal Copenhagen)

are hand-painted. Those that are to be painted over the glaze (Fig. 7.7) are dipped in the appropriate glaze and then go into the glost kiln for a second firing. The under-glaze painted wares are glazed over the enamel and then fired, of course.

As was explained in Chapter 6 only a limited number of coloured enamels will stand up to the high temperature of the glost kiln, so the more colourful figurines are painted over the glaze and the enamel fired at lower temperatures. Many people have a preference for the subtly shaded under-glaze painted figurines like those produced by Royal Copenhagen, and it is the limited palette dictated by the under-glaze technique which gives these figurines

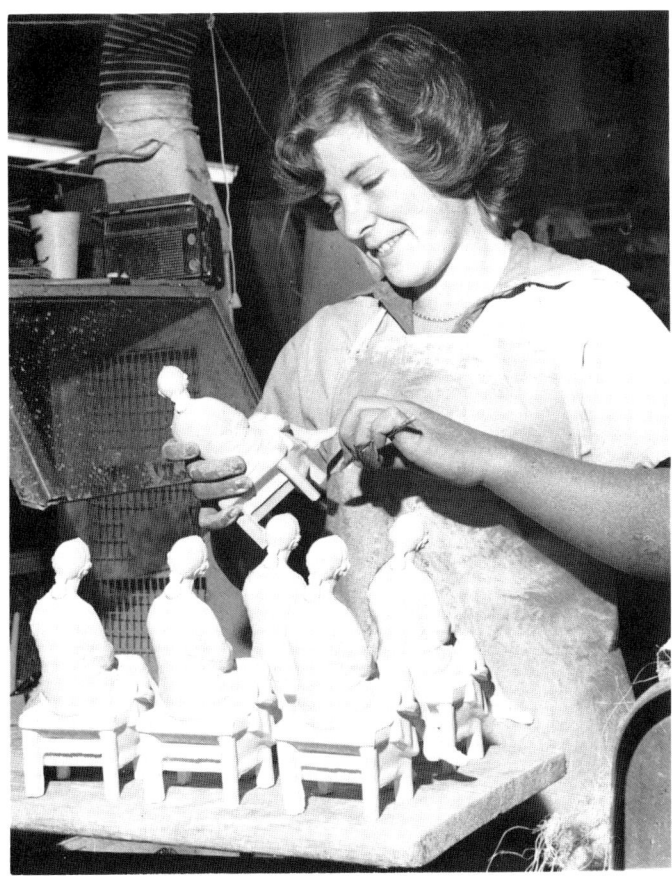

Fig. 7.6. Fettling a figurine at the Royal Doulton factory

their immediately recognizable character. Others, however, show a marked preference for the brighter colours possible with over-glaze painting, and, of course, if designers want to render birds and flowers in a naturalistic style then they have no alternative but to opt for painting over the glaze.

Fig. 7.7. Painting Royal Doulton Figureheads in enamels over the glaze

Painting

The painting of figurines is very labour-intensive. At Royal Doulton no fewer than 260 painters are employed on this operation. After painting and firing, a factory which produces fine figurines will again submit every one to another searching inspection. A common fault is to find minute specks of dirt trapped in the glaze. Often these can be ground out by a skilled craftsman, but this may, of course, involve reglazing or repainting. Just what can be involved in ensuring that every figurine is perfect is summed up by one of the inspectors at Royal Doulton: 'One of our figurines may go through as many as six firings before it is best, and this factory won't turn out anything but best.' This attitude is equally true of all the leading factories, and it is one of the arguments the salesman can put forward to justify the price of quality products. Put another way, it is a factor which the pottery buyer should bear in mind when he is selecting stock. He should appraise samples and deliveries with the same critical eye which the inspector in a reputable factory brings to this job. Only then can he be sure he is buying 'best'.

8 Advice on caring for pottery and glass

Everyone selling china and glassware should be able to advise customers about how to take care of their purchases in order to protect them from damage or breakage. They should also be in a position to tell the public how to deal safely with any problems that may arise, such as the tarnishing of gold decoration.

Looking after bone china and porcelain

Bone china and porcelain are very strong materials, and the wares made from them will stand up to a good deal of wear and tear. It is, however, possible for pottery to be damaged or broken if some simple precautions are not taken to protect it. The main threats to these wares are:

(1) Abrasion
(2) Thermal shock
(3) Mechanical shock
(4) Acids, detergents and soap

Abrasion

It is not generally appreciated that glazes can be scratched. Most people would probably realize that steel wool, pan scrapers and soap pads should never be used for cleaning fine china, but what many people do not know is that it is other pieces of china which are responsible for most of the scratches that appear on plates and cups. Just as diamond cuts diamond, so does glaze scratch glaze. To drop one plate on another, or to scrape one across another, can easily lead to scratching, and this is most likely to happen when these wares are wet. Tests carried out by the manufacturers have revealed that most scratching occurs when plates are stacked while they are still wet in draining racks and are accidentally rubbed one against the other in the process. So china should be handled carefully, both in the washing-up water and during subsequent stacking. Another common cause of scratching results from wearing rings when handling china, and it should be appreciated that diamond is not the only gemstone that can gouge the surface of glass or glaze. Porcelains are less likely to be scratched in this way than bone china, as the hardness of porcelain is 8 on Mohs' scale and therefore only ruby, sapphire and diamond are likely to scratch it. Needless to say, great care should be taken when loading china into a dishwasher to ensure that the pieces do not touch one another. Care should also be taken in storing pieces after use. It is a good idea to place a piece of paper towelling between plates to prevent the foot of one plate scratching the top surface of the one below it. Cups should never be stacked one inside another.

Heat

A bone china or porcelain cup will never crack as a result of pouring boiling water into it unless it has already been cracked or subjected to mechanical shock. But heat can damage china. Though porcelain is better able to withstand heat shocks than bone china, it is not sensible to put expensive porcelain plates in an oven to heat them, unless they are specially designed to be oven-to-table wares. They can be heated over a stove or in a compartment of a cooker designed for the purpose, but probably the safest way to heat fine china is in hot water. It is not advisable to use water at too high a temperature in a dishwasher, not because this will result in breakages due to heat shock but because chemical attack is stimulated by high temperatures, and this may result in damage to the decoration on china.

Mechanical shock

Most breakages result from mechanical shock. Even if a plate does not break immediately as a result of being dropped on the floor or given a hard knock, it may well do so later when subjected to heat, so every care should be taken to save china from hard knocks. A rubber mat in the sink and rubber nozzles on the taps can prevent many breakages. This is another reason why cups should never be stacked one inside another. This practice results in the damage resulting from one knock being multiplied, and can also lead to a weakening of the handles.

Acids, detergents and soap

Acids, such as fruit juices and vinegar, can attack the decoration on china, and these substances should not be left on plates or in bowls but rinsed off as soon as possible. If china is left to soak for long periods in a detergent solution, that is, in washing-up water, this can cause glaze to lose its shine or coloured enamels to fade. Soap solutions are even more likely to cause this kind of damage. In this context soap is not so kind and gentle.

Another problem with detergents is that they can form a scum on the surface of a ware which mars its appearance. This is not likely to happen with hand-washing, but it is quite likely to build up in a dishwasher. If the scum does not respond to subsequent hand-washing it can be removed relatively easily by polishing with a silver plate powder.

Gold problems

Gold decoration is more liable to damage resulting from wear than is enamel decoration. This is because gold cannot be fired into a glaze as completely as can an enamel. So particular care should be taken to protect china decorated with gold from abrasion during washing-up and in storing it. It is also very important to avoid using abrasive powders for washing-up such wares.

If wares decorated with gold are stored for long periods the gold may acquire a bloom or may even tarnish. This can be removed either with talcum powder and a damp cloth or with a dilute ammonia solution.

Silver marks

Because glaze is harder than silver, silver or silver-plated utensils can be abraded by china, and this results in traces of silver appearing on the glaze. It can normally be removed in the course of washing-up, but if the marks prove stubborn recourse may have to be made to silver plate powder.

Tea stains

Some substances can stain a glaze and the worst culprit is unquestionably tea. Tea stains can be quite stubborn, particularly if a glaze has been scratched or if a detergent scum has formed, in the crevices of a teapot, for example. Polishing with a damp cloth using either common salt or borax will usually shift any stubborn stains, and cotton buds are useful for getting into odd corners. After cleaning, care should be taken to rinse a ware cleaned in this way. Borax is poisonous, and salt does not improve the taste of tea.

Oven-to-table wares

While many oven-to-table wares such as oven-proof porcelains can be taken out of a deep freeze and used immediately for cooking, these wares, when hot, should not be placed in a refrigerator until they have been allowed to cool slowly.

The care of glasswares

Glasswares, if they are to be protected from damage and maintained in mint condition, also need careful handling. They may not be as fragile as they look, but they will, of course, break or chip if they are dropped onto a hard surface or if they are knocked. It is not always realized that, even if they appear to have survived an accident, they may break subsequently, seemingly coming apart in the hand during washing-up. This is because the blow they had suffered set up stresses that resulted in cracking in the warm water. So, like pottery, glassware should always be carefully handled and carefully stacked, and again a rubber mat in the sink and rubber nozzles on the taps can prevent many breakages.

Heat

Glass is even more likely than pottery to break as a result of heat shock. Perhaps it goes without saying that only glasswares specifically designed for the purpose should be used for serving hot beverages. Customers may not, however, be aware that glass should be washed up in water at a low temperature. Lead crystal should be washed up in water that is only hand-warm and other glasswares should not be placed in water at a temperature higher than 60 degrees Centigrade. People who use dishwashers need to be especially careful to ensure that this temperature is not exceeded. When hand-washing glasses they should be put in the bowl one at a time not only to ensure against a build-up of heat, but also to prevent one ware being banged against another, and glasses should be dried immediately after washing and not left to drain.

Water

Water is, rather surprisingly, one of the worst enemies of glasswares. This is because water absorbs alkali from the glass, taking in traces of the soda or potash which are among the constituents of the metal. These alkalis combine with the carbonic acid in the atmosphere to produce carbonate of soda or carbonate of potash, which appears on the surface of the glass as a white deposit. This can be washed off, but if it is allowed to build up it will gradually destroy the finish and dull the glass. It follows that glasswares should be stored in a dry atmosphere and never in the packing in which they come from the shop. This packing tends to hold moisture and destructive acids, and if glasswares were stored in it for a long period chemical erosion could take place that would turn the glass iridescent. This iridescent appearance can be seen on ancient glasswares that have been in the ground for a long period and so suffered from this chemical action.

Sunlight

It is not wise to leave glass standing in sunlight for any length of time. While sunlight will not affect lead crystal, it will turn soda glass yellow. Some soda glass does have a tendency to yellow with age, and sunlight seems to accelerate this process.

9 The industry today

The collection of profiles of firms incorporated in this chapter, while it includes many of the best-known names, is far from being comprehensive. There are many hundreds of glasshouses and potteries, large and small, in Europe and it would obviously be quite impossible to include every one of them. These profiles nevertheless serve to provide a cross-section of the two industries, illustrating both the diversity and the variety of their products.

Baccarat

Baccarat is in Lorraine, France, and was at one time part of the estates of the Bishop of Metz. It was one of the bishops of this diocese, Monseigneur de Montmorency-Laval, who petitioned Louix XV in 1764 and successfully persuaded him to authorize a glasshouse in the region to make good use of the vast forests in the area. He argued that France produced no art glass and 'this is why Bohemian glassware is imported in such large quantities, causing an enormous outflowing of funds at a time when the kingdom has such great need of them to recover from the terrible seven years' war'. The glasshouse that was set up as a result of this petition has survived three revolutions and four invasions.

Initially the firm had only three furnaces, but between 1766 and 1789 it had consumed 600 tons of sand and 300 tons of rock salt, and was burning 6400 cords of timber a year to fire its furnaces. It began by producing glasswares in the Venetian style and by 1817 was also making lead crystal.

Antoine Renaut, who had been engaged to run the factory by the bishop, retired in 1802 and left the works to his three sons. But the revolution and the war that followed left the firm in financial difficulties, and in 1806 the factory was sold. It was eventually acquired by the owner of a crystal factory in Belgium, a M. d'Artigues, who again sold the company to three of his associates, and in 1824 the company we know today, the Compaignie des Cristalleries de Baccarat, came into existence. From this time the factory produced the wares which it still specializes in – stemwares, toilet articles, perfume bottles, decorative wares, candelabra and chandeliers – and in 1846 started to produce the millefiori paperweights for which it became famous and which are today eminently collectable.

In 1824 a glass-maker named Robinet devised the pump, which bears his name, for blowing large and heavy pieces, and at this time the firm had a hundred craftsmen working in their cutting shop. In 1836 another shop was set up and the workforce doubled. Baccarat also became famous for their wheel engraving at this period. One of their artists, Simon, gained a worldwide reputation for his skill. Subsequently the technique of acid-etching, using hydrofluoric acid, resulted in a less expensive form of decoration. In 1867 the firm installed the first gas-fired furnace. In 1880 the cutting of glassware by machine was adopted and Baccarat were the pioneers of this technique.

The factory was occupied by the Germans during the First World War and was a prisoner-of-war camp for a time under German occupation during the Second World War. After 1945 the company concentrated on the American market, setting up an office in New York in 1949. Besides the cut stemware patterns such as Harcourt (Fig.9.1), which they produced for the United States and their other export markets, they also gained a reputation, particularly in France, for their plain stemware patterns such as Brummel and Perfection. The production of their plain patterns was made possible by the introduction of continuous electrically and gas-heated furnaces, which they began to install in 1965, and which were the first of their kind in Europe. They say that 'these furnaces produce night and day the most perfect lead crystal of optical quality'. Today the factory employs 1200 people and, besides stemwares, produces chandeliers and decorative

Fig. 9.1. Baccarat's Harcourt cut stemware (by courtesy of Baccarat)

wares. Among these are the Béstiaire Fabuleux collection, produced in collaboration with the famous jewellery house of Chaumet. For some time Baccarat have carved animal heads from blocks of glass. The new collection consists of animal figurines made from carved glass and gold and set with a variety of gem material.

Bohemian glass

The present Czechoslovakian state-owned glass industry is both large and diverse. It is comprised of big firms and small, and of artist-craftsmen working out of their own studios. Many of the firms have their own specialities and the combined industry produces every imaginable type of useful and ornamental glassware.

Czechoslovakia consists of three distinct regions, Bohemia, Moravia and Slovakia, and glasshouses exist in all three. The most famous centres of the industry are, however, Nový Bor and Zeleznéy Brod, but other important factories are located at Harrachov and at Moser in Bohemia.

The industry employs 5000 people, produces 120 000 different items of glassware and exports to eighty countries. It produces hand-blown and machine-made glass and uses a number of different types of metal. The industry manufactures a lead crystal, which, like most of the lead crystal made on the Continent, contains 24 per cent of lead. It makes a number of types of flint and soda glasses and also a semi-lead glass, which contains 10 per cent of red lead and has a markedly better colour than flint glass.

Glass-making in Bohemia probably dates from the eighth or ninth centuries, as a glass furnace dating from the ninth century has been found at Nitria, but production at that time was probably restricted to beads. From the twelfth century there is documentary evidence that stained window glass was being produced for the

Fig. 9.2. These cased glass bowls are made from ruby and colourless glass in one of the traditional Egermann designs from Czechoslovakia. The red glass is etched away to produce the design (by courtesy of the Nový Bor Glassworks)

monasteries and glasswares were produced in this area certainly as early as the fourteenth century. From the sixteenth century onwards the Bohemian glass-makers produced exquisite engraved glasswares, the finest of these being made in Silesia in the first half of the eighteenth century.

Supplies of Czech glasswares to the retail trade are channelled through Vitrea, who act as commission agents for Glass Export. The designs produced range from undecorated stemwares that are similar to Scandinavian designs to the famous heavy lace-cutting, which is a speciality of the Bohemian glassworks at Poděbrady. Some Czech glasswares are very decorative, like the rich red cased wares first introduced by F. Egermann in 1829 and, nowadays, produced at the Nový Bor glassworks (Fig. 9.2). This factory also produces glasswares painted in enamels, while Moser Karlovy Vary make suites of stemwares which combine cutting and lace-like gold decorations.

Royal Copenhagen

The Danish royal family were largely responsible for setting up The Danish Porcelain Manufactory in 1775. The chief benefactress was Queen Juliane Marie, who was motivated partly by a desire to improve the national economy by reducing imports of porcelains from Germany and partly by family rivalry. Her brother, Duke Charles I of Brunswick, had set up the Fürstenburg factory in 1753, and she felt that Denmark could be just as successful in producing fine porcelains. Initially, it did not seem that the enterprise would, in fact, be successful. Indeed, by 1779 many of the shareholders had become so restive that King Christian VII was persuaded to acquire all the shares and become sole owner. On 30 June of that year the firm was rechristened The Royal Danish Porcelain Manufactory. By this time the factory had accumulated stock of 100 000 pieces, and a sale was organized in 1780, but customers were few, averaging only three a day. Most of the pieces produced at this time were vases, a number of which are still in the royal palaces in Copenhagen, as well as tea and coffee services. After 1781 the factory also began to produce complete dinner services. Most of these and the other wares were bought by the royal family for their own use or as gifts.

In 1790 work began on a dinner service for King Christian VII, which was to become one of the world's most famous patterns and is still in production to this day. A copy of a botanical work, *Flora Danica*, was sent from the royal palace to the factory, the idea being that the beautiful plates in the book should form the decoration for a royal dinner service, each piece to be decorated with a different example of Scandinavian flora. Two years later the service, designed for eighty people, was delivered, 988 pieces in all. By 1794 another 363 pieces were added, and further additions were made in subsequent years. In 1862 'a committee of ladies was formed' in Denmark with the object of presenting Princess

Fig. 9.3. Royal Copenhagen's Flora Danica pattern (by courtesy of Royal Copenhagen)

Alexandra with a Flora Danica service on the occasion of her marriage to the future King Edward VII of England, and this the second service to be produced is now kept at Sandringham (Fig 9.3).

Another of Royal Copenhagen's famous patterns, the underglaze painted blue and white pattern, known in England as Blue Fluted, also dates from the eighteenth century, though in the 1880s many new pieces were added and modifications made to the pattern by the famous designer Arnold Krog.

During the past century Royal Copenhagen have also become famous for their porcelain figurines, the most characteristic of which are painted under the glaze. In recent years they have also given considerable encouragement to studio potters, who work either in the factory or in their own workshops, and produce stonewares which are displayed in the firm's showrooms and shops alongside the factory's porcelains.

Today the factory produces 5 000 000 pieces a year bearing their famous three-wave mark, symbolizing the three waterways that dominate Denmark. They employ 1600 people, 600 of whom are painters. The firm established an English branch in 1897.

The Dema Glass Group

This group consists of Dema Glass, who specialize in one-piece machine-made stemwares and produce 100 million glasses a year, and two producers of hand-made cut crystal wares, Thomas Webb and Edinburgh Crystal.

Thomas Webb

This firm owes its foundation to John Webb who became associated in the 1830s with the White House Glassworks at Wordsley in Staffordshire. Thomas Webb the first had been involved in a partnership in another glasshouse, the Wordsley Flint Glassworks, but when he inherited the White House in 1835 he withdrew from Webb and Richardson. Shortly after this he also withdrew from the White House to found his own firm, Thomas Webb and Sons, and from 1837 to 1856 operated Platts

Glasshouse, located at Amblecote. In 1856 he opened a new glasshouse nearby, known as Dennis Glassworks. He was succeeded in the 1860s by his three sons, Thomas Wilkes Webb, Charles Webb and Walter Wilkes Webb. In 1919 the Dennis Glassworks was acquired by Webb Crystal Glass, and in 1964 the two firms became part of Crown House Ltd, of which the Dema Group is a division. Thomas Webb still make fine cut crystal. One of the firm's specialities are the long-necked decanters they have been producing for over a hundred years.

Edinburgh Crystal

This firm also make fine cut lead crystal and in its range are a number of distinctly Scottish designs, the best-known of which is their thistle pattern. The firm dates back to the seventeenth century, when glass-making began in Leith. By the nineteenth century there were a number of glasshouses in this area, among them the Norton Park Glassworks of Jenkinson & Co. The firm changed its name in 1866 to the Edinburgh and Leith Flint Glass Company. In 1919 it was acquired by Thomas Webb, and in 1955 its name was changed to the present one. Together with Thomas Webb, it became part of the Dema Group in 1964, and ten years later the Norton Park glasshouse was closed and the firm moved into a modernized factory at Penicuik, near Edinburgh.

Royal Doulton

Today this is a very large group producing a wide range of products sold under different brand names. Besides the Doulton factories the group also owns the Minton and Crown Derby factories, Paragon, Royal Albert, Beswick, Colclough, Royal Adderley Floral, Booths and Ridgway, Dunn Bennett and the Webb, Corbett glass factory.

Royal Doulton

In 1815 John Doulton, having served an apprenticeship at John Dwight's Fulham Pottery, invested his entire savings of £100

($150) to buy a partnership in a pottery at Lambeth. His son Henry Doulton succeeded him at Lambeth, and besides pioneering stoneware drainpipes in 1846 he also became increasingly interested in producing better designed decorative wares, establishing a studio and exploiting the talents of leading artists of the day. In 1877 he bought a factory in the potteries, in Nile Street, Burslem. Ten years later he was knighted by Queen Victoria, the first potter to receive this honour. Then in the reign of Edward VII he was granted the Royal Warrant and was authorized to add the word Royal to his company's name.

Initially the Burslem factories produced lead-glazed earthenwares, but in 1884 the manager of the factory, John C. Bailey, persuaded Doulton to let him introduce bone china.

It was Charles Noke who laid the foundations of Doulton's reputation for producing figurines. He modelled the first in the 1890s, and after his appointment as the factory's art director in 1909 he commissioned well-known sculptors to model figures which the factory cast in bone china. These included Charles Vyse, who modelled the Darling figurine which the firm still produces. In the 1920s and 1930s the range was greatly expanded, and also in

Fig. 9.4. Royal Doulton's Image range in white bone china (by courtesy of Royal Doulton)

the 1930s they revived an old Staffordshire tradition and began to produce Toby and character jugs. In 1934 a nun sketched rabbits on scraps of paper and these sketches were translated into finished designs at the outbreak of the Second World War by Hubert Light. There are now 150 pieces of this Bunnykins ware for children and they have proved exceptionally popular. The firm's vast range of tablewares and figurines includes both bone china wares and wares made from the English translucent china (Fig. 9.4) introduced by the then-technical director, Richard Bailey, who subsequently became the firm's managing director. This relatively inexpensive translucent body won the firm a Queen's Award for Technical Innovation in 1966. The firm has also more recently introduced a range of oven-to-table wares called Lambeth.

Minton

This factory, founded in 1793 by the freelance designer and engraver, Thomas Minton, became part of the Royal Doulton Group in 1968. The business was initially a partnership between Thomas Minton, William Pownall, who provided the finance, and the brothers Joseph Poulson, a master potter, and Samuel Poulson, a model-maker, with Thomas Minton's brother acting as the firm's agent in London.

The firm began by producing blue-printed earthenwares, mostly of the Willow Pattern, but by 1805 they were already producing bone china. Thomas Minton died in 1836 and was succeeded by his son, Herbert Minton, whose success soon attracted artists from other companies. In 1848 the experienced French potter, Léon Arnoux, came to Minton, and introduced maiolica wares into the firm's range. Another Frenchman, Louis Marc Solon, joined the company in 1870. He came from the Sèvres factory, and brought with him a skill in the delicate *pâte-sur-pâte* technique of decoration. Meanwhile, in 1863 James Leigh Hughes had developed the acid gold technique and the Minton factory subsequently acquired the patent rights for this process. Fifteen years before they became members of the Royal Doulton Group, Minton had built a new factory in London Road, Stoke-on-Trent. Today the factory produces bone china tablewares and decorative wares, hand-painted or transfer-printed. They also specialize in acid gold decoration and in gold-paste decoration.

Royal Crown Derby

This factory became part of the Royal Doulton Group in 1964 and its characteristic gold and red patterns of the Imari type remain as popular as ever (Fig. 9.5).

There is no record of the date when pottery was first produced in Derby, but it is believed that a Huguenot goldsmith's apprentice, Andrew Planche, persuaded the Derby banker John Heath to join him in setting up a factory there early in the 1750s. It was in this factory near St Mary's Bridge that Planche was producing 'dry edge' figurines as early as 1752. There is also a record of three small cream jugs which date from this period.

William Duesbury, an enameller, joined the partnership in 1756 and a sizable factory was opened on Nottingham Road. Following this, the firm expanded into tablewares, mainly blue and white wares in the chinoiserie style with the decoration painted over the glaze in Smith's Blue, named after Constantine Smith, the factory's head colour-man. William Duesbury purchased the famous Chelsea factory in 1769 and opened a London showroom in 1773. He closed the Chelsea factory in 1784, having in the meantime transferred many of the skilled craftsmen to Derby. In the following decade the factory was to become one of the famous potteries in Europe under the management of William Duesbury II, who gathered round him a group of talented decorators and made improvements both to the bodies and the glazes. One of the decorators was Michael Kean, an Irish painter of miniatures, who became a partner in the firm in 1795 and took over after Duesbury's early death in 1797. Unfortunately, under his management the quality of the wares declined, and the family forced him to sell out. Then in 1811 Robert Bloor leased and subsequently bought the factory. It was under his management that the Japanese Imari patterns were developed, though the Imari patterns, in fact, date back to 1770. By the time of Bloor's death in 1846, the factory was again in financial difficulties, and was sold two years later to Alderman Clarke, a relation by marriage. A group of employees, however, opened a factory at King Street,

Fig. 9.5. One of Royal Crown Derby's Imari patterns (by courtesy of Royal Crown Derby)

and they produced both figurines and Imari pattern tablewares. In 1877 a new china factory was opened in Derby by two men from the Worcester factory, Edward Phillips and William Litherland. The new firm was named The Derby Crown Porcelain Company, and the factory was built in Osmaston Road. This factory again attracted many talented artists, including a number who came over from the Continent at the end of the Franco–Prussian war. The factory became famous not only for the superlative decoration these artists created but for a new 'eggshell' body of incredible fineness and for raised gold and pierced designs. In 1890 the company was appointed Manufacturers of Porcelain to Her Majesty, which meant they could add Royal to their title. It merged with the King Street factory, which traded under the name Sampson Hancock, in 1935.

Royal Albert

This firm was founded by the Wild family. The first member of the family to be involved in the pottery industry was Isaac Wild, who became a pot-handler in Stoke-on-Trent. His son Thomas followed his father into the trade, became a saucer-maker and rose to be a manager. In 1896 Thomas Wild took over the Albert Works in Longton and was joined in this enterprise by his son Thomas Clarke Wild, who had started in the trade at the age of eight as a jigger-turner and had eventually gone into business for himself. In 1898 Thomas Clarke Wild inherited the firm on the death of his father, and at this time the factory became known as Royal Albert Crown China Works. It specialized in bone china tea and breakfast sets, which were sold at very competitive prices. The firm employed 120 people and was so successful that they bought St Mary's Works from the Moore Brothers in 1906 to enable them to increase their production capacity. Three years later they had to expand this factory to cope with demand, and acquired other factories between 1910 and 1920 as a result of their considerable success in both the export and home markets. The firm began to move up-market and introduced a number of innovations into their factory. They installed the first electric Rotelec Kiln in 1937 and the first twin continuous china biscuit ovens in 1946. They became part of the Royal Doulton Group in 1964 and demand for their traditional designs such as Old Country Roses has led to the factory increasing its production considerably in recent years.

Beswick

This factory, which was founded in Stoke-on-Trent in 1896 and became part of Royal Doulton in 1969, specializes in animal figurines, ranging from portraits of famous racehorses to Beatrix Potter characters.

The founders of the firm were James Wright Beswick and his son, who began making tablewares and ornaments, but later the firm gained a reputation for producing the accurate models of horses and domestic pets for which it is so well known.

Royal Adderley Floral

This factory at Longton specializes in a branch of the potter's craft which only dates back to the 1920s. Although hand-modelled flowers were added to figurines and tablewares long before this, floral pieces that existed in their own right date back only some sixty years. At this factory the flowers are modelled by hand, petal by petal, then fired, glazed and hand-painted.

Colclough

This company came into the Royal Doulton Group after a merger with S. Pearson and Sons in 1972. (Pearson had acquired Colclough in 1944.) The factory was founded by H. J. Colclough in Longton in the last century and sold by his son to Booths Limited. The firm has pioneered mechanization and rationalization in order to produce bone china wares at a moderate price, and in the 1960s introduced a range of replaceable bone china designed to persuade the public to treat bone china as an everyday commodity.

Booths and Ridgway

This factory was also acquired in the merger with Pearson. They make fine earthenware, including the famous Brown Betty teapots and Willow Pattern tablewares.

Paragon

The Star China Company was owned by Herbert Aynsley at the beginning of the twentieth century and produced fine china tea and breakfast services. In 1919 the name of the firm was changed to Paragon. When Herbert Aynsley retired in 1927 he was succeeded by Hugh Irving, and under him the firm received a royal appointment from Queen Mary. The firm have produced a loving cup to celebrate every royal occasion since the 1920s. Today they produce a range of tablewares and fancies.

Webb, Corbett Ltd

This old-established firm became part of the Royal Doulton Group in 1969 and the products of the factory, cut lead crystal wares, are now sold under the Doulton name. These include, besides table glass, a large range of 'fancies' such as bud vases and thimbles.

Webb, Corbett also originated in the White House Glassworks with which the Webb family had been associated as early as 1833. This was sold by the Richardsons, who had been associated with the Webb family in business from the outset, to the grandson of Thomas Webb the first in 1897. It was at this time that the firm Thomas Webb and Corbett Ltd came into existence. In 1913 the old glasshouse was destroyed by fire and another glasshouse was taken over by the firm at Coalbourn Hill, north of Stourbridge. Meanwhile in 1906 the firm had taken over a glasshouse at Tutbury in Staffordshire. In 1913 the firm changed its name to Webb, Corbett Ltd.

Royal Grafton

The Grafton China Works in Stoke-on-Trent came into being in 1876, being founded by Alfred B. Jones. In 1900 he took his two sons into partnership and moved his operation into a new factory at Longton in Staffordshire, and it is here that the firm is still located. The Joneses specialized in bone china tea and breakfast sets. Then in 1906 they diversified into heraldic china, with which they were very successful until the demand for this fell off in the 1920s. In 1956 they added a range of dinnerwares and these, their tea sets and a range of giftwares are what they produce today. In 1971 the firm was acquired by the New Zealand-based firm, Ceramco.

Holmegaard

Kastrup and Holmegaards Glassworks is the most famous of the Danish glasshouses. The firm owes its foundation to Countess Henriette Danneskiold-Samosøe, who started the Holmegaards Glasvaerk in 1825. The Count, her husband, had planned to start a glassworks to exploit the plentiful supplies of peat on Holmegaards Moor in Zeeland, but he died in 1823 before he could realize his plan. The factory made green glass bottles for Denmark's breweries, and in 1847 a second glassworks was set up at Kastrup, near Copenhagen. Then the industry was consolidated with the Royal Copenhagen Porcelain Company in 1924. In 1925 Jacob E. Bang joined the firm as designer, and it was he who established the firm's reputation for table glasswares and decorative wares of simple and vigorous form. Jacob Bang's son, Michael Bang, and the firm's present designer, Per Lütken, carried on the tradition, bringing their own skills to the creation of new forms. In 1954 the Holmegaard and Kastrup factories were amalgamated.

Lalique

This famous French glass firm was started by the leading figure in the *art nouveau* movement in France, René Jules Lalique. Besides designing decorative glasswares and jewellery he designed perfume bottles for Coty, a very early example of the use of recognizably designed packaging in the cosmetic industry. His factory, started in 1908, was located at Combs, near Paris. He was a follower of Emile Gallé in carving cased glass in the *art nouveau* style and also experimented with sand-blasting and acid-etching to achieve unusual surface textures. In 1918 he acquired the glasshouse at Wingen-sur-Moder in the Bas-Rhin department, which is now run by his son Marc Lalique under the name Cristallerie Lalique et Cie.

Nuutajarvi Glass

Glass has been produced in the village of Nuutajarvi, north of Helsinki, in Finland since 1793. Glass-making is still carried on in the old glasshouse there by Oy Wartsila AB. The range, like that of other Scandinavian glasshouses, has been created by a number of industrial designers and much of the production is plain glass, but the glasshouse also produces cut glass, coloured glass and even latticinio designs. Heikki Orvela designed a latticinio range called Filigranai in 1981.

Orrefors

Orrefors is located in the Småland region of Sweden and was originally an iron works, but in 1898 it was converted to a glasshouse for making bottles for medicine and ink, window glass and inexpensive domestic glasswares. Then in 1913 Johan Ekman acquired it and began to produce decorative glasswares. This was the period in which so many Scandinavian factories producing consumer durables called in the services of designers, and the Orrefors factory was no exception. Simon Gate and Edward Hald were the first of a succession of distinguished designers employed by the factory, and it was Hald who extended the carving technique evolved by Emile Gallé to make Graal glass. This was carved cased glass, which was reheated to soften the outlines and then covered with a gather of clear glass. He also began to evolve the new decorative techniques for which the factory became famous. Among these was Ariel glass, a technique consisting of entrapping air bubbles to create patterns in the glass which is associated with the designer Vicke Lindstrand, who worked for Orrefors Glassbuk from 1928 to 1941. Then there was Kraka glass, associated with the designer Sven Palmqvist, and named after a lady from the Viking sagas who appeared to her lover dressed in a fishing net. This type of glassware consists of a network of opaline or coloured glass between layers of clear glass.

In 1946 the company was bought by Henning Beyer, and is perhaps best known today for its plain stemwares and decorative wares, which are greatly admired. The glasshouses also produce engraved glasswares, cased glass and chandeliers.

Poole

This firm dates back to 1854, when a pottery manufacturer from Hanley in Staffordshire, John Ridgway, and a group of local businessmen established The Architectural Pottery Company at Hamworthy, near Poole. Seven years later, in 1861, the firm's chief technician James Walker broke away and started his own firm. Walker, however, soon went bankrupt, and his business was purchased by Jesse Carter, a successful ironmonger and builder's merchant from Weybridge in Surrey. He operated the business so successfully that in 1895 he was able to buy The Architectural Pottery as well. Jesse Carter's firm continued to make architectural tiles, and their only venture into pot-making was to produce large garden pots for Liberty and Co. in the early part of the twentieth century. As early as 1904, though, Jesse Carter's son, Owen Carter, began producing vases, probably just as a hobby, and some of these can be seen in the collection at the factory. Owen Carter was very interested in the arts and was involved with Roger Fry's Omega Workshops during the First World War. It was probably as a result of his interest in the potter's craft and the enthusiasm of the head of the firm's design department, James Radley Young, which led the company to begin to produce pottery between 1914 and 1918. In a report for the year 1920 it was stated that 'We are proud of our pots. Started by Mr Young when things were slack during the war, they have met with a very encouraging reception by the discriminating public, with the result that what started more or less as an experiment has now become an established, though comparatively small factor in our business, and Carter pots are to be found all over England, and have spread to America'.

During the 1920s the firm continued to expand its pottery production, introducing a range of hand-thrown and hand-painted wares designed by Truda Carter. It was during the 1930s, however, that the firm first gained their reputation for domestic pottery. Owen Carter had brought the pottery designer John Adams and the silversmith Harold Stabler into the firm in the 1920s, and it was these two artists who now really established the image of the pottery. However, Truda Carter still continued to make an important contribution. She was responsible for the colour combinations which became a recognizable characteristic of the wares. Some of the best-known designs from this period were John Adams' Sherborne and Streamline patterns.

Though the official name of the pottery was Carter, Stabler and Adams, it soon became known to the public as Poole, and in 1952 the factory began to mark its wares 'Poole, England'. Then ten years later the name of the pottery was changed to Poole Pottery. In 1971 the pottery became part of the Thomas Tilling Group.

Today the firm employs 275 people in its factory on the quay at Poole. It produces both earthenwares and stonewares (Fig. 9.6)

Fig. 9.6. Poole's Kismet range of oven-to-tableware (by courtesy of Poole Pottery Ltd)

and has recently added a range of bone china gift wares. The factory utilizes modern production machinery, including a roller machine that can turn out 4000 plates a day.

Rosenthal and Thomas

There are a large number of potteries in both East and West Germany, including, of course, the famous Meissen factory at Dresden. Probably the best known of all these firms in the UK today is, however, the Rosenthal Group, which includes the Thomas factories. Like most Continental firms, this group specializes in porcelain and it also has a flourishing glass factory.

Thomas

Founded at the beginning of the twentieth century, this firm became part of the Rosenthal Group in 1908. The factories of Porzellanfabrik F. Thomas are located at Kulm and at Werk Waldershof, near Bayreuth, and employ upwards of a thousand people. The firm also has a glass-works at Amberg in the Upper Palatinate, which was designed by Walter Gropius the founder of the Bauhaus, whose belief it was that 'form should follow function'. This dramatic building is known in West Germany as 'The Glass Cathedral'. The porcelain tablewares and the glasswares produced by this firm show that it is very design-conscious, and Thomas currently employ the services of such distinguished artists as Hertha Bengtson from Sweden and Robin Levien from England.

Rosenthal

Rosenthal were already well established by the time they acquired Thomas in the first decade of the twentieth century. The founder, Philip Rosenthal, emigrated to the United States in 1872 and worked as a dishwasher and a cowboy before becoming the manager of a New York firm of pottery and glass importers. In 1879 he returned to Germany, and set up in business for himself at Selb, where the firm still has its headquarters. He worked out of a cellar of Erkersreuth Schloss and employed one assistant, a porcelain-painter. Initially he purchased surplus saucers from other manufacturers, decorated them and sold them with no small success in the American market. Eventually he began to produce his own blanks, and from the outset he was concerned to produce original designs. One of his most successful early designs was the *art nouveau* pattern, Donathello, introduced in 1906. He lived to be eighty-two, dying in 1937.

Fig. 9.7. The Gropius teaset from Rosenthal, offering an ideal combination of function and aesthetics (by courtesy of Rosenthal)

His son, Philip Rosenthal junior, did not join the firm until 1950. He had originally wanted to become a university don, but war intervened. He joined the French Foreign Legion, was a prisoner of the Vichy regime and eventually escaped to England via Gibraltar. When he joined the family firm he was appointed advertising manager, and, after taking responsibility first for product design and then for sales, he became chairman in 1958.

If it was his father who laid the foundations of the company, it has been Philip Rosenthal junior who has expanded it to its present prominence. In 1971 he founded Rosenthal Technik, which has since became the largest producer of industrial ceramics in Europe, and during his time with the company the group's turnover has increased twelve times. His interest in design is demonstrated by the fact that he is the president of the German Design Council, but it is also clearly apparent from the product range which the firm produce. He is on record as saying that 'we don't believe in following the market, but in shaping it', and to this end he has employed the services of many of the leading industrial designers of our times. Among these have been Walter Gropius (Fig. 9.7), Tapio Wirkkala from Finland and Lord Queensberry from England. Henry Moore and Victor Pasmore have been among those who have designed 'art pieces', while Salvador Dali and Eduardo Paolozzi are two of the famous artists who have created decoration or designed wares for the company.

Besides the standard lines which Rosenthal produce in their factory at Selb, incidentally also designed by Walter Gropius, they produce a studio line. These limited editions, some functional and some purely decorative, are designed for them by artists from half a dozen different countries.

Rye Pottery

This pottery, although a small operation, is interesting, because since it was reopened after the Second World War by the brothers John and Walter Cole, it has specialized in maiolica wares. Also known as English Delftwares, these are produced by painting the decoration onto the glaze before it is fired, so that the colour becomes incorporated in the glaze and acquires a characteristic softness. This pottery was originally started in 1869, an offshoot of the Cadborough Farm Brickworks, which was operated by the Mitchell family. They had already achieved the distinction of supplying the Prince of Wales and the Queen of Spain with 'rustic fancy ware' as early as 1866. The old factory was known for a ware called 'sprigged green hops' and the firm today employs a hop motif on their 'Hop Ware'.

Stevens and Williams (Royal Brierley)

This glass factory has direct links with the Henzey family, one of the families which Jean Carré brought over to England from Lorraine in the seventeenth century. The brand-name under which the firm sells its glasswares, Royal Brierley, derives from the location of the factory. The Moor Lane glasshouse was situated on what used to be called Briarlea Hill. It was owned by a Mr Honeybourne, and in 1740 his daughter married John Pidcock, who was a nephew of the last surviving member of the Henzey

Fig. 9.8. A water set in Royal Brierley's Braemar design (by courtesy of Stevens and Williams (Royal Brierley))

family. In the nineteenth century the glasshouse was leased by a Joseph Silvers, whose daughters married William Stevens and Samuel Cox Williams, who became principals of the firm in 1847 and changed its name to Stevens and Williams Ltd. The firm is still in the hands of the Williams family. The present chairman is Lieutenant-Colonel R. S. Williams Thomas, the fifth generation of the family to be associated with the company, and his two sons are currently active in the firm. They produce a wide range of cut crystal, intaglio-cut and engraved glass (Fig. 9.8).

Stuart Crystal

In 1827 an orphan of eleven years of age, Frederick Stuart, went to work at the Red House Glassworks at Wordsley in Staffordshire. He subsequently joined another Wordsley firm, John Parrish and Co., for whom he worked as a traveller. In 1856 he went into partnership with Richard Mills and a member of the Webb family, and as Mills, Webb and Stuart this partnership operated the Albert Glassworks in Wordsley. Then in 1881 he acquired the glasshouse where he had gone to work as a boy.

The Red House Glassworks, where Stuart Crystal is still located, was erected shortly after the cutting of the adjacent canal which could carry the coal from the Staffordshire mines to fire the furnaces and the clay deposits to make the pots. It was these mines and these deposits which attracted the industry to the area. It is probable that the 260-metre (85-foot) cone-shaped glasshouse, recently restored for conversion into a working museum, dates from the 1780s or the early 1790s.

Frederick Stuart lived on until 1900, dying at the age of eighty-three, and the firm still remains in the hands of the Stuart family. The present chairman is Ian Stuart, a great-great-grandson of the founder of the firm.

In 1915 the firm acquired the White House Glassworks across the road from the Red House and today the Vine Street factory, rebuilt on this site, has a total area of 7300 square metres (24 000 square feet). The firm have also set up a factory in South Wales at Aberbargoed, which now has nine electrically heated single-pot furnaces and employs 120 people. In the main factory over a hundred glass-cutters are employed.

Stuart Crystal 161

Fig. 9.9. The Ariel pattern, by Stuart, features an air-twist (by courtesy of Stuart and Sons Ltd)

The firm's range includes not only cut crystal but intaglio-cut designs, stemwares featuring mixed intaglio and cut decoration, engraved glass and wares decorated by sand-blasting. They also have in their range a pattern with an air-twist stem (Fig 9.9). In recent years Stuart have also set up a retail division and today operate twenty-six shops within shops.

Waterford

A glasshouse existed in Waterford, Ireland, as early as 1729, but it seems to have been short-lived. Then in 1783 George and William Penrose petitioned Parliament for assistance in setting up a flint

Fig. 9.10. One of Waterford's famous ship's decanters (by courtesy of Waterford Crystal Ltd)

glassworks. The Penroses were assisted in founding this enterprise by John Hill, a Stourbridge glass-maker who brought over a number of skilled workmen to Ireland. The fact that there was no excise duty on glass in Ireland encouraged the making of substantial wares suitable for cutting, and the Waterford factory soon gained a reputation for its cut crystal wares. A duty was introduced in 1825, however, and by 1851 the factory, like many other Irish factories, found itself unable to compete with the English glass-makers, who by then had been relieved of the burden of a duty.

Glass-making came back to Waterford in a small way in 1947, but the pilot glasshouse set up by the Irish wholesaler Bernard Fitzpatrick was taken over in 1950 by Joseph McGrath, chairman of the Irish Glass Bottle Company. He brought over glass-blowers from West Germany and Italy and engravers and cutters from both these countries and from Czechoslovakia to set standards and train Irish labour. By 1967 the firm had outgrown the glasshouse in which it was set up, and a new factory was built. Since then, Waterford has become the largest producer of cut crystal in the world. The factory is most famous for its diamond faceting and its hob-nail cuts (Fig. 9.10).

The Wedgwood Group

This group today consists of no fewer than ten pottery firms and two glass-manufacturing plants, and they recorded sales of £117 million ($175 million) for the year ending in April 1982. Besides the modern Wedgwood factory at Barlaston, which employs 2000 people and which attracts no fewer than 100 000 visitors a year, the group includes such household names as Coalport, William Adams, Mason's Ironstone and Crown Staffordshire. The group operates twenty factories, employs 8000 people and produces 20 per cent of all the ceramic tablewares in the UK.

Josiah Wedgwood was born into a family of potters. He set up his own firm in 1759 and soon after he opened his factory he produced the cream earthenwares with a transparent glaze, which were purchased by Queen Charlotte in 1765 and became known as Queen's Ware. He introduced his jasper wares in 1774 and continued to improve them in subsequent years. He died rich and famous in 1795.

The firm began to make bone china in 1812 under Josiah Wedgwood II, but production ceased as a result of the slump in demand during the Napoleonic Wars, but was revived in 1878. The firm still produce jasper wares, decorated with cameo reliefs, and

Fig. 9.11. The Florentine design by Wedgwood, hand-enamelled on bone china. The original pattern was designed over a century ago (by courtesy of the Wedgwood Group)

the black basalt wares that were first made in 1768. They make a wide range of bone china wares both hand-painted, hand-gilded and decorated by the lithographic process (Fig. 9.11). They also produce the oven-to-table wares first introduced into the range in 1965, though the firm had produced culinary wares as early as the 1790s.

Coalport

Coalport was founded in 1750 and later became famous for its fine bone china wares under John Rose, who purchased the factory in 1799. One of their achievements was the introduction of a leadless feldspathic glaze, which won an award from the Society of Arts in 1820. By the time Wedgwood took them over in 1967 the firm had three factories in Stoke-on-Trent.

William Adams

William Adams was one of the earliest of the Staffordshire potteries. It dates from 1657, when John Adams rebuilt Brick

House, Burslem, as a pottery. They joined Wedgwood in 1966 and are one of the leading earthenware manufacturers in the UK.

Mason's Ironstone

Ironstone was a body containing a slag of ironstone, invented by Charles James Mason, which he patented in 1813. Mason was the son of Miles Mason, who had started a factory at Lane Delph near Newcastle-under-Lyme in 1813. The firm failed, and was purchased by Francis Morley, who was to make a flourishing business out of selling ironstone wares. The firm was sold in 1859 and again in 1881, when it was acquired by Goddards, who eventually sold it to Wedgwood in 1973. Today the factory in Hanley produces both tablewares and ornamental wares in earthenware, and the original ironstone shapes and patterns are still being produced.

Other firms in the group

Johnson Brothers, founded in 1883, eventually had three factories in Hanley producing earthenware tablewares, and became part of Wedgwood in 1968.

Crown Staffordshire was founded in 1810 and was one of the first firms to produce ornamental bone china floral pieces. They also produce bird models, giftwares and traditional tableware patterns. This factory, at Fenton, Stoke-on-Trent, was acquired by Wedgwood in 1973.

J & G Meakin, founded in 1851, produced tablewares in earthenware at their Hanley factory. This firm merged with the Burslem firm of Midwinter, which produces earthenware, oven-to-table wares and stonewares, in 1968. Both factories became part of Wedgwood in 1970.

The two other pottery factories are Enoch Wedgwood (Tunstall) Ltd, founded in 1835, which made earthenwares in Stoke-on-Trent and joined Wedgwood in 1979, and an American firm, Franciscan, in California, which make earthenwares and architectural tiles at Glendale. They became part of the group in 1979.

Fig. 9.12. A lead crystal decanter from Dartington (by courtesy of the Wedgwood Group)

Wedgwood Glass and Dartington

One of the glasshouses is Wedgwood Glass, which became part of the group in 1969. This was formerly King's Lynn Glass Ltd, founded in 1966. The other is Dartington, which the firm acquired in 1983. The Dartington glass works was established in 1966 at Dartington Hall at Torrington, near Totnes in North Devon. Eskil Vilhelmsson was appointed managing director and brought over skilled workers from his native Sweden to start production and train local labour. Although Frank Thrower was responsible for the designs and won a Duke of Edinburgh Design Prize for his work in 1972, the glasswares produced by Dartington have a marked Scandinavian flavour (Fig. 9.12). The factory produces hand-made lead crystal glasswares and today employs 200 people and exports to fifty-three countries.

Royal Worcester

This company owns four factories today. The factory at Worcester was established in 1751 and therefore is the oldest surviving firm manufacturing porcelains in England. Spode was founded by Josiah Spode in 1770, when he took over the factory of Turner and Banks, of which he had previously been the manager. Then there are the Palissy and Hammersley factories.

Worcester

The Worcester factory was one of three established within a few years of one another with the intention of producing wares that would rival those imported from China and Germany. The other two were the Bow and Chelsea factories. Bow and Chelsea both produced soft-paste porcelain, but Worcester made a porcelain similar to that produced in China.

A brass and copper manufacturer, Benjamin Lund, is credited with being the instigator of hard-paste porcelain when he obtained a license to explore for feldspathic rock in Cornwall. It seems, however, that it was a Plymouth apothecary, William Cookworthy, who eventually opened a pottery factory in Bristol in 1749, who first produced this porcelain. Cookworthy's firm was acquired by the newly formed Worcester Porcelain Company two years later. Two of the minor shareholders in this venture were Dr John Wall and William Davis, and they transferred the workforce from the Bristol factory to Worcester, including the secrets of porcelain-making. It is thought that they were given their shares in exchange for this valuable arcanum. The factory's early wares are often referred to as belonging to the Wall period, and it seems that he was responsible for perfecting the manufacture of the porcelain which had still been in the experimental stage at Bristol. There is no doubt that this porcelain gave Worcester a distinct advantage over her rivals, a fact which Robert Dossie was to note in his *The Handmaid to the Arts*, in which he wrote that 'a manufactory at Worcester has produced, even at very cheap prices, pieces that . . . bear hot water without more hazard than the true China ware'; something which could not, of course, be said of the soft-paste porcelains from Bow and Chelsea. Because of the resemblance of the Worcester porcelain to the Chinese, it was inevitable that the factory should produce chinoiseries, and the most famous pieces from the eighteenth century are the under-glaze painted blue and white wares in the Chinese style.

Worcester also commercially exploited transfer printing during the eighteenth century, first printing over the glaze but subsequently using under-glaze printed decoration. Within a decade of its foundation the factory was also producing wares decorated over the glaze in enamel and applying brilliant ground colours, among them the lovely green used at Sèvres and a pink scale ground borrowed from Meissen. The factory also produced figurines in a small way.

In 1783 the firm's London agent, Thomas Flight, bought the factory and in 1793 took Martin Barr into partnership. This resulted in that confusing description of Worcester wares, between 1783 and 1840 as Flight, Flight & Barr; Barr, Flight & Barr; and Flight, Barr & Barr. The firm became a limited company in 1862 under R. W. Binns, and for the first time the word Royal was now incorporated in its title because of the royal patronage it was receiving.

From 1912 to 1953 Charles William Dyson Perrins was in charge of the factory. He was a collector of early Worcester pieces, and his collection forms a major part of the contents of the museum at

Fig. 9.13. Pottery figurines modelled by Dorothy Doughty for Royal Worcester (by courtesy of the Dyson Perrins Museum, Worcester)

the factory which bears his name. Today the firm produces both fine bone china and porcelain wares of many shapes, decorated in a variety of ways, including the use of acid gold. They are also famous for their figurines, which include those beautiful birds which were modelled for the factory by Dorothy Doughty and produced in limited editions (Fig. 9.13), and Doris Lindner's equestrian pieces. Another of the firm's specialities are the decorative pieces hand-painted with fruit and other motifs in lavish detail.

Spode

The Spode factory also dates back to the eighteenth century. Josiah Spode had begun his career in the pottery industry in Stoke-on-Trent at an early age, having been sent to work at the age of seven, and he had risen to be a factory manager under William Banks before acquiring the firm of Turner & Banks. He went on to develop all-over transfer printing under the glaze and was experimenting with a new body incorporating bone ash when he died in 1797. Three years later this Spode body, which the factory still produce today, was introduced by his son, also called Josiah Spode. In 1812 the firm also introduced a body closely resembling porcelain, and in 1821 they put a feldspar porcelain on the market. Josiah Spode had a partner, William Copeland, who was based in London, and in 1829 his son, William Taylor Copeland, became the sole proprietor following the death of the founder's grandson only three years after he joined the company.

In 1842 Copeland introduced Parian wares and in 1856 he was the first pottery manufacturer to install the filter press invented by Needham and Kite. In 1868, the year before his death, his sons joined him in the business and the family remained associated with Spode until 1931.

Today Spode still produce bone china. They make stoneware which dates back to 1813 and earthenwares including the Indian Tree pattern which they first made in 1878. They make a range of modern figurines designed by Pauline Shone, miniature wares which they first made in 1815, presentation plates and ceramic handled cutlery.

Two other potteries are associated with Royal Worcester; Palissy, who have recently begun to produce bone china, and Hammersley, who produce Victoriana and gift lines in bone china.

Select Bibliography

An Illustrated History of Glass, H. Newman, Thames and Hudson, 1977
An Illustrated Dictionary of Ceramics, G. Savage, Thames and Hudson, 1976
The Arthur Negus Guide to British Glass, J. Brooks, Hamlyn, 1981
The Arthur Negus Guide to English Pottery and Porcelain, B. Price, Hamlyn, 1978
Pottery and Porcelain Figures, T. Hughes, Country Life Books, 1981
Directory of European Porcelain, L. Danckert, NAG Press Limited, 1981
Royal Doulton Figures, Eyles and Dennis, Royal Doulton, 1978
Meissen Porcelain in Colour, H. M. Fletcher, Exeter Books New York, 1979
Eighteenth Century Wedgwood, D. Buten, Pitman House, 1980
The Royal Copenhagen Porcelain Manufactory 1775–1975, Royal Copenhagen, 1975
Worcester Porcelain, S. Fisher, Ward Lock, Undated
French Porcelain, W. B. Honey, Faber and Faber, 1950/1972
German Porcelain, W. B. Honey, Faber and Faber, 1947
English Pottery and Porcelain, W. B. Honey, A. and C. Black, 1933/1969
Victorian Pottery, H. Wakefield, Herbert Jenkins, 1962
Japanese Porcelain, S. Jenyns, Faber and Faber, 1965
English Blue and White Porcelain of the Eighteenth Century, B. Watney, Faber and Faber, 1963/1973

English and Irish Glass, G. Wills, Guinness Signatures, 1968
Old Glass, O. N. Wilkinson, Ernest Benn Ltd., 1968
Continental Glass, F. Davis, Arthur Barker, 1972
The Book of Glass, G. Weiss, Barrie and Jenkins, 1971
Glass, R. H. Vose, The Connoisseur, 1975
The Chinese Potter, M. Medley, Phaidon, 1976
Pottery Science, M. Wickham, Pitman/Watson–Guptill, 1978
Pottery, H. Hodges, Hamlyn, 1972

Glossaries

Glass

ACID-POLISHING After glass is cut the facets have to be polished. This is done by acid-etching in a bath containing hydrofluoric and sulphuric acid.

AGATE GLASS Glass of several colours combined to resemble the microcrystalline form of quartz which is called agate.

AIR-TWIST A form of stem decoration common in the second half of the eighteenth century and still used today. It is produced by inducing a bubble in the metal by pegging it and then drawing out the metal and twisting it.

ANNEALING Taking out the stresses in glasswares by passing them through a lehr.

BALUSTER STEM A type of stem which was first used in the seventeenth century. It was derived from the balustrades of Renaissance architecture.

BATCH The mixture of constituents which are put into the pot to fire.

BLISTERS Air bubbles, resulting from gases created during the melting of the constituents of the metal not being driven off.

BLOWING The expansion of the metal taken from the furnace by the craftsman blowing through his hollow blowing iron. Glass can either be free-blown, blown in a mould by a foot-blower or mechanically in an automated machine.

BLOWING-IRON The 2-metre (6-foot) long hollow steel tube on which the foot-blower gathers his metal, marvers it and then blows it.

CAMEO GLASS A cased glass which has been carved to resemble a cameo. The most famous example of cameo glass is the Portland Vase, made in Rome and now in the British Museum.

CANE A rod of glass made by combining different metals. Millefiori decoration is produced from canes.

CASED GLASS Glass consisting of a second gather of a different colour surrounding the initial gather. Cameo glass is produced by carving the outer layer of cased glass.

CHAIR The group of men who make glasswares, for example, a wine chair consists of four men. The leader of the group is called the gaffer. Also the crude wooden seat with extended arms on which the gaffer sits.

CLAW BEAKER A type of sixteenth-century German glassware in which the punts have been blown out and extended to resemble claws.

COLOURED GLASS Glass may be coloured by the addition of metallic oxides. Cobalt oxide produces blue glass, copper oxide green glass, etc.

CORDS Lines, or streaks, in the glass resulting from poor mixing of the constituents or from the joining of metals of different temperatures.

CORE TECHNIQUE The way the earliest glasswares were made by dipping a rod, on which a sand or clay core had been built up, into molten glass and then marvering it. Afterwards the core was removed.

CRACKING-OFF Removing the moil by scribing the ware with a diamond tool and then passing it through a gas flame. Nowadays the process has been automated.

CRIMPER A tool used for crimping the rim of a bowl.

CRISTALLO A type of glass first produced in Venice. It was almost colourless as a result of great care to ensure that the constituents were uncontaminated and of the use of decolourizers.

CRIZZLE A disease of glass resulting in internal cracking, the formation of a sour-smelling liquid and eventual disintegration. Probably the result of an excess of alkali. Ravenscroft's lead crystal proved not to suffer from crizzling.

Glass 175

CULLET   Broken glass from the shrower which is put in the pot and melted with the other constituents. It assists fusing, but is also included for economic reasons. Makers of fine glasswares use only their own cullets.

CUT GLASS   Glass that has been faceted by grinding, first on a carborundum wheel and then on a sandstone wheel.

DECOLOURIZERS   Chemicals added to the constituents of glass to neutralize any metals present which might otherwise tint the glass.

DIPSTICK   A rod used to gather metal to produce trailed decoration.

DISEASED GLASS   See CRIZZLE

DUCTILE   Easily drawn into a desired shape.

EDGE-MELTING   Melting the rim of a glass in a gas flame, after the moil has been cracked-off, to round the sharp edges. Nowadays the edges are usually ground in automated machinery.

ENAMELLING   Decorating glasswares with enamels consisting of a frit of metallic oxides and glass. After this has been painted on the glass it is fired.

ENCRUSTED GLASS   Wares from which small fragments of contrasting coloured glass project, having been attached after marvering.

ENGRAVED GLASS   Glass can be engraved in different ways, with a diamond tool or under a small copper wheel charged with carborundum powder and oil.

ETCHING   Decorating glass by attacking it with hydrofluoric acid. The areas which are not to be etched are protected with a wax resist and the ware submerged in a bath of the acid.

FAÇON DE VENISE   In the Venetian style. Glass-makers all over Europe copied the delicate cristallo wares made on the island of Morano for many centuries.

FIRING   The process of fusing the enamel or gold decoration on a glassware.

FLASHER   Another name for a siden, so called because it flashes as it is revolved on a turntable.

FLINT GLASS   Glass made from ground-down flint pebbles instead of silica sand. Flint is an impure form of quartz and therefore largely consists of silica dioxide.

FLUTE-CUTTING Cutting a series of grooves on glass that lie parallel to one another or which fan out. There are flat flutes, such as those applied to stems, and hollow flutes, applied to the bodies of vessels.

FOOT-BLOWER The member of the chair who gathers the glass from the pot and blows it.

FRIT Fused glass which is subsequently ground to a powder. Mixed with metallic oxides, this process produces enamels.

FROSTED FINISH Glass which has not been polished after cutting. The sandstone wheel leaves the glass with an opaque white surface.

GAFFER The leader of the chair who forms the stems and feet of stemwares.

GATHER The metal which the foot-blower or the bit gatherer takes from the pot on his blowing iron or punty iron.

GILDING The application of gold to glass. Mercurial gilding was the technique used at one time, the gold being amalgamated with mercury and then the mercury vapourized. Nowadays gold is fired onto the glass.

GLORY HOLE The small furnace located in the caterhole shop where larger wares are produced. These wares may have to be reheated a number of times as the workman forms them and they are placed in the glory hole for this purpose.

HOBNAIL-CUTTING A style of cutting that produces small circular mounds or points. A style popular in Ireland and nowadays associated with the Waterford factory.

HYDROFLUORIC ACID An acid which readily attacks silicates and which is therefore used both for etching and polishing glasswares.

ICE GLASS Glass which resembles ice. It can either be produced by plunging the hot metal into cold water and then reheating, or by covering the marvered metal with small splinters of glass and reheating.

INTAGLIO-CUTTING A method of cutting carried out below a small sandstone wheel, giving far more freedom to the cutter than cutting over a larger wheel. Intaglio-cutting is often left frosted, but is sometimes polished.

LATTICINO Clear glass embedded with threads of contrasting coloured glass, usually opaque white.

LEAD CRYSTAL   Glass containing red lead. In the UK there is a British Standard for lead crystal which stipulates that it must contain 30 per cent lead. On the Continent most glasshouses produce a lead glass containing 24 per cent lead.

LEHR   The tunnel oven through which wares pass on a moving belt to remove the stresses in the glass. The wares are first raised to working temperature then slowly cooled.

MARVER   A steel sheet, or recessed block of wood, on which the foot-blower rolls the metal on his blowing-iron to ensure it is true. Also the act of marvering.

METAL   The molten glass taken from the pot in the furnace.

MOIL   The end of the sphere attached to the blowing iron after blowing that is cracked-off before the ware goes into the lehr.

MOULDED GLASS   Wares produced by pouring or injecting molten metal into a mould.

NAILSEA GLASS   Wares made in the glassworks at Nailsea, but also a term applied to parti-coloured wares produced in various glasshouses in the Bristol area.

POT   The ceramic vessel in which the constituents of glass are melted in the furnace.

PUNTY IRON   Sometimes called a pontil iron. This is the solid iron rod on which the gathers are taken from the pot to form the stem and foot of a wine glass. Also the iron attached to the base of wares in the caterhole shop after they have been taken off the blowing-iron.

SAND-BLASTING   Engraving glass by directing a jet of sand onto the surface. The parts of the glass which are not to be engraved are protected by a rubberized resist. Used mainly to engrave crests on glasswares.

SEEDS   Small bubbles in glass due to the gases formed during the fusion of the metal not being completely driven off.

SEMI-LEAD CRYSTAL   Glass containing a small amount of lead, perhaps 10 or 12 per cent.

SERVITOR   Another name for the gaffer who occupies the chair in a wine chair.

SHROWER   The bin into which broken vessels from the lehr and the moils from the blowing irons are put to be later used as cullet.

SIDEN   A ware which is out of true.

SODA LIME GLASS Glass whose constituents are soda and lime and silica sand. Soda lime glass contains no lead.

TEARDROP An air bubble induced, in the metal used to form the stem of a stemware, by pegging. This decorative effect was common in the baluster stems of the early eighteenth century and is still occasionally to be seen in modern glasswares.

TRAIL DECORATION Trailing molten glass from a dipstick or heating molten rods of glass and trailing them to produce decorative effects.

WOB FOOT A ware whose foot is out of true.

WOODS The wooden boards used by the gaffer to form the foot of a stemware.

WORKMAN The gaffer in a caterhole shop

Pottery

ACID GOLD A process of producing intaglio gold decoration by etching the glaze with hydrofluoric acid and then covering the incised area with gold.

BALL CLAY A type of clay found in Devon and Dorset which is cut into balls.

BASALT A black stoneware produced by Josiah Wedgwood from Staffordshire clay coloured by coal from nearby seams.

BAT A slab of body placed on the jigger to make a plate or on the jolly to produce a cup.

BISCUIT Pottery which has a matt surface after firing. Hence a biscuit kiln and biscuit firing.

BLACK PORCELAIN Sometimes called *porcelaine noire*, coloured by metallic oxides and specially fired.

BLEEDING The spreading of enamel during firing that causes one colour to run into another.

BODY The prepared ingredients from which articles of pottery are made.

BONE The calcined cattle bones which are ground down and used to produce bone china.

BONE CHINA A type of pottery made from 50 per cent bone ash, 25 per cent china clay and 25 per cent Cornish stone.

BORDER   The rim of a plate or vessel. Many descriptive terms are used to describe the various types of border, such as egg and tongue, ribbon and leaf, husk, etc.

CALCINE   To reduce bone or stone with intense heat so that they can readily be reduced to powder.

CAMEO   Relief decoration of the kind applied to Wedgwood jasper wares.

CASTING   Pouring slip into a plaster of Paris mould to produce hollowares or figurines.

CERAMIC   A generic term applied to all forms of fired clay (derived from the Greek *Keramikos*, meaning pottery).

CLAY   Decomposed granite consisting of aluminium oxide, silica dioxide and water. When the water is driven off by heat this plastic material becomes permanently hard.

COBALT BLUE   The oxide of cobalt which, when mixed with a powdered glass, produces the blue enamel used to decorate pottery. It was used by the Chinese, and later by the European potters, for under-glaze painting because of its resistance to high temperatures.

COILING   Making pottery by coiling long sausages of clay to form the walls of pots.

COLOURED BODIES   These are produced either by natural impurities in the clay or by additives in the form of metallic oxides. Jasper wares are examples of bodies coloured by metallic oxides.

CORNISH STONE   A feldspathic stone found in Cornwall which is similar to the China stone used by the Chinese potters to produce porcelain.

CRAZING   Fine cracks in the glaze caused by shrinkage during the cooling of the glaze. Can be caused by placing wares in too hot an oven.

CREAM WARES   Wares produced from a cream body. A well-known example was the Queen's Ware produced by Josiah Wedgwood.

DELFTWARE   A tin-glazed earthenware made in Delft, Holland. The term 'English Delftware' is used to describe similar wares made in England.

DEMI-TASSE   A small coffee cup, popular in the USA.

DIP-DECORATED   Wares decorated by dipping in a coloured slip.

DRESDEN CHINA   A name applied to Meissen porcelain, but also used for other wares made in that city and not infrequently to figurines in the Meissen style made elsewhere.

EARTHENWARE   Low-fired clay which remains porous unless covered with a glaze. Terracotta is an unglazed earthenware. Faience is a tin-glazed earthenware.

EGGSHELL   A very thin ware of porcelain or bone china. Originally produced in the Ming period in China and later produced by a number of English factories.

ENAMEL   Metallic oxides mixed with ground glass and suspended in an oil. Enamels are painted or printed onto pottery and then fired to fuse them into the surface.

ENCRUSTED   Decorated with relief motifs such as applied flowers.

FAIENCE   A tin-glazed form of earthenware sometimes known as Maiolica or Delftware.

FAMILLE ROSE   Pink decoration. *Famille verte* is green decoration, *famille jaune* is yellow and *famille noire* black.

FELDSPAR   A rock largely composed of aluminium silicate which acts as a temper for kaolinite clay. It fuses to form a glassy substance which binds the kaolinite particles. Used in bodies and glazes.

FETTLE   To trim excess material from a piece of pottery before firing. The fettling tool is a small-bladed scalpel.

FIRE CLAY   Clay capable of withstanding high temperatures used to produce kiln furniture.

FIRE-SPECKLING   A defect in pottery caused by alien black material such as carbon in the body or in the glaze.

FLUTED   A ware decorated with grooves imparted by a mould.

FOLIATE   Decorated with a leaf pattern.

FOOT   The base on which a vessel stands.

FRIT   A fused mixture of glass and enamel subsequently ground down to a fine powder.

GILDING   Applying gold which is subsequently fired.

GLAZE   A glass-like substance applied to a body after it has been fired. Glazes are used to make earthenwares waterproof, to decorate stonewares and to improve appearance and produce the translucency of bone china and porcelain.

GLOST FURNACE   The furnace in which a glaze is fused onto the body of pottery wares.

GRINDING   Removing glaze or the marks of the supporting furniture from the foot of a ware.

GROUND   A monochrome area of colour.

HARD-PASTE PORCELAIN   The type of porcelain produced by the Chinese potters and copied in Europe in the eighteenth century. Made from kaolin-rich clay and feldspathic stone, it is hard and strong (c.f. SOFT-PASTE PORCELAIN).

HIGH-FIRED   Fired at high temperatures. Stonewares are high-fired as compared with earthenwares.

HOLLOWARES   Wares of hollow form such as cups, jugs, pots and vases as compared with plates and dishes, which are flatwares.

HYDROFLUORIC ACID   The only acid which will attack silica products such as pottery and glass. Used in the acid gold process.

IMARI   A type of Japanese pottery, the decoration on which derives from Japanese textiles. It was widely copied in Europe, and Imari designs were, and still are, a speciality of the Royal Derby factory.

IMPRESSED   Indented as a result of stamping the body before firing. Some factories employ an impressed distinguishing mark.

INCISED   Scratched. Some factories use incised indentifying marks.

IN-GLAZE DECORATION   Decoration applied to a glazed ware. The ware is then fired at a temperature which causes the glaze to melt and absorb the colour. Colours are restricted to those enamels which will withstand the high temperature of the kiln.

INTAGLIO   Incised decoration acid-etched, or carved, into the surface of the body.

IRONSTONE   A type of pottery containing slag of ironstone produced by Charles James Mason after 1813.

JIGGER   The semi-automatic machine used for making a plate. It consists of a powered turntable on which a mould is placed and a profiling tool that shapes the base of the plate.

JOLLY   A semi-automatic machine similar to the jigger, which is used for making cups.

KAOLINITE   Aluminium silicate, the major constituent of kaolin, the white clay used to produce porcelain and bone china.

KILN An alternative name for furnace, e.g. biscuit kiln, glost kiln.

KILN FURNITURE The supports for wares in the furnace. they are made from various heat-resistant substances, including fire clay.

LEAD GLAZE One of the earliest forms of glaze used for pottery, containing lead oxide.

LITHOGRAPHIC PRINTING The application of enamel decoration printed on a plastic sheet either by the lithographic or the silk screen process. This decoration is usually polychromatic. Subsequent firing fixes the enamels and burns off the plastic sheet.

LUSTREWARES Wares with a metallic surface.

MAIOLICA (OR MAJOLICA) Named after the island now known as Majorca, this was originally a tin-glazed earthenware made in Italy, but the word was later used to describe similar earthenwares made in England.

MARKS Over the centuries potters have marked their wares in various ways, using painted, incised or impressed marks. A number of guides to antique pottery marks are available.

MODELLING Shaping the original model for a ware or a figurine. The modeller usually works in clay with spatulas.

MONOCHROME Of one colour only, i.e. monochrome decoration.

MOULD The hollow plaster of Paris block into which slip is poured to produce hollowares and figurines.

OVER-GLAZE DECORATION Enamels painted or printed onto the ware after it has been glazed.

PAINTED DECORATION Enamel decoration applied by hand with a brush.

PARIAN WARE An unglazed porcelain, introduced by Copeland, which resembles marble and was used to produce figurines.

PASTE An alternative word for body, often used to describe a porcelain body.

PASTE GOLD A variation on *PÂTE-SUR-PÂTE*. The applied paste decoration is enriched with gold.

PÂTE-SUR-PÂTE Paste-on-paste. Body applied to body and carved to produce delicate patterns. The technique was first used in China. It was adopted by the Sèvres factory in the nineteenth century and later in the the century by the Minton factory.

PÂTE TENDRE   A name sometimes used to describe soft-paste porcelain because of its delicacy and fragility.

PETUNTSE   The Europeanized version of the Chinese word for the feldspathic rock used to produce porcelain.

PIECE MOULD   The mould used to produce a part of a ware or a figurine. Most items are produced in a number of different moulds and the parts later assembled.

PIERCED DECORATION   Decoration produced by cutting holes in the body before it is placed in the biscuit furnace for firing.

PITCHER MOULD   A recessed mould into which body is pressed to produce a casting. The cameo decoration on jasper wares is cast in pitcher moulds.

PLASTICITY   Of a nature which allows it to be readily modelled. Clay has considerable plasticity.

POLYCHROME   Many coloured, as in polychromatic decoration.

PORCELAIN   A translucent body evolved by the Chinese, the secret of which was discovered at Meissen in the early eighteenth century. It consists of a clay largely composed of kaolinite and feldspathic stone, nowadays usually with the addition of quartz.

POTSHERD   A fragment of pottery. Potsherds found on historic sites were invaluable aids to dating before the development of the carbon-dating technique which archaeologists now employ.

POTTER'S WHEEL   The horizontal wheel on which a potter throws his pots. Sometimes operated by a foot-treadle but nowadays usually mechanically powered.

POTTERY   A generic word to cover all articles made from clay, including porcelain, bone china, stoneware and earthenware.

PRINTED DECORATION   Decoration produced by applying an enamel, printed on tissue paper from a copper plate. The tissue paper is sponged away and the enamel which has been transferred to the pottery is then fired. Usually this decoration is monochrome.

PUGGING   The process of kneading the body in a purpose-built machine to ensure a smooth mixture with the correct water content.

REDUCING ATMOSPHERE   An atmosphere in a kiln which is rich in carbon monoxide. It affects the colour produced by metallic oxides. Copper oxide will produce a bluish-red colour in a reducing furnace.

REFRACTORY   A term used to describe clays or clay products that will resist high temperatures.

RELIEF DECORATION   Decoration that stands out, in relief, from the background.

RESIST   A protective material. Used in the acid gold process to restrict the etching effect of the acid to the required areas.

SAGGER   A protective casing of fire clay used to contain wares during firing in the kiln.

SALT-GLAZED   Glazed by throwing common salt into a glost kiln. The sodium in the salt combines with the silicates in the body to form a glass-like skin over the ware.

SHRINKAGE   All pottery shrinks in the biscuit kiln. The degree of shrinkage depends on the nature of the body, ranging from 2 to 20 per cent.

SLAB   A technique sometimes used by craft potters to produce pots. It consists of assembling slabs of body. Slab pots tend to be square in form.

SLIP   Body in a liquid state used for pouring into moulds to cast wares. Also used to decorate earthenwares and stonewares by applying after the wares have been formed.

SOAPSTONE   Another name for Cornish stone, the feldspathic rock used in the production of porcelain.

SOFT-PASTE PORCELAIN   Also known as artificial porcelain, *porcelaine de France* and *pâte tendre*. Produced by mixing clay and a glass frit, it is softer than true porcelain. It can be scratched with a file and also has poor resistance to thermal shock.

SPONGE   To smooth the surface of pottery with a damp sponge before firing.

STONEWARE   A ware which is part-way between an earthenware and a porcelain. It consists of a mixture of clay and a fusible stone and is fired at a high temperature, which results in its being impervious to water.

TERRACOTTA   Unglazed earthenware of an orange-brown colour, used mainly for garden ornaments.

THROWING The process of fashioning a ware on the potter's wheel.

TIN GLAZE A form of glaze incorporating tin oxide used to produce maiolica, faience or Delftware.

TOBY JUG A jug in the form of a seated figure in eighteenth-century costume. Probably first made by Ralph Wood in the second half of the eighteenth century.

TOW To polish the surface of pottery with a fibre mop, usually made of flax but sometimes of hemp or jute.

TRANSLUCENT Allowing the passage of diffused light. Bone china is the most translucent form of pottery.

TURNING Decorating a pot on the wheel with a chisel in the same way that a metal-worker turns a piece of metal on a lathe.

UNDER-GLAZE DECORATION Enamel applied to the biscuit body before the glaze is applied and fired. A technique developed by the Chinese potters and widely used in Europe in the eighteenth century. Colours are restricted to those enamels which will withstand the temperature of the glost kiln.

VITREOUS Glassy. Glazes give pottery a vitreous surface.

WASTER A ware which is damaged or distorted in the kiln and is therefore unusable.

Index

Abrasion, 132, 134
Acceptance, 39–41
Acid effects, 134
Acid-etching, 152
Acid geld, 85, 107, 115–116, 147, 169
Acid-polishing, 51–52
Ackerman, John, 18
Adam brothers, 76, 80
Adams, John, 155, 164
Adams, William, 164
Agate glass, 23
Agates, 6, 115
Air-twists, 19, 20, 46, 161
A la Mode, 122
Albert Glassworks, 160
Albrechtsburg Schloss, 70
Alexandra, Princess, 143
Amphorae, 61
Anglo-Saxon glass, 8
Annealing furnace, 19
Antique glassware, 21
Arcanum, 71, 75
Architectural Pottery Company, 154
Argentine, 93
Arial, 46, 153, 161
Arneux, Léon, 146
Arsenic, 26, 27, 28
Art nouveau, 23, 152, 157
Assembly, 100
Athens, 64
Aynsley, Herbert, 151

Babylon, 64
Baccarat, 137–139

Bailey, John C., 145
Bailey, Richard, 146
Baluster stem, 12, 19, 45–47
Bang, Jacob, 152
Bang, Michael, 152
Banks, William, 169
Barium sulphate, 79
Barlaston, 163
Barr, Martin, 168
Bas-Rhin, 153
Bats, 100
Battan, Thomas, 82
Battersea Enamel Works, 110
Belleek Pottery, 121
Bengtson, Hertha, 156
Bentley, Thomas, 77
Béstiaire Fabuleux, 139
Beswick, James Wright, 150
Beyer, Henning, 154
Binns, R. W., 168
Biscuit-firing, 100–101
Bit-gatherer, 31, 36
Black basalt ware, 78, 87
Blisters, 39, 42, 52
Bloor, Robert, 147
Blue below the glaze, 68–70
Blue Flowers, 108, 109
Blue Fluted, 99, 107, 108, 109, 143
Blue glass, 6, 25, 45
Bodmin Moor, 86
Body, 87–88, 90, 91
Bohemia, 12, 23, 139–141
Bone, 93
Bone ash, 73, 90, 170

Bone china, 73, 90–93, 100, 103, 121, 122, 150, 151, 156, 164, 169, 170
  caring for, 132–136
Booths and Ridgway, 150
Booths Limited, 150
Borax, 26
Böttger, Johann Friedrich, 70, 121
Bow, 73, 90, 121, 167
Boy with Calf, 124, 125, 126
Braemar, 159
Bristol, 13, 75, 168
Brooks, John, 110
Brown Betty teapots, 150
Brummel, 138
Brunswick black, 116
Brunswick, Duke of, 71, 141
Bunnykins ware, 146

Cadborough Farm Brickworks, 159
Calcium antimony compound, 45
Calver, Valerie, 62
Cameos, 6, 23, 79
Campania, 6
Carborundum wheel, 49
Carré, Jean, 13, 159
Carter, Jesse, 154
Carter, Owen, 154, 155
Carter, Stabler and Adams, 155
Carter, Truda, 155
Carving, 76
Cased glassware, 23, 45
Castings, 99
Ceramco, 152
Ceramics, 87, 158
Chair, 31
Charles I, Duke of Brunswick, 71, 141
Charlotte, Queen, 163
Chaumet, 139
Chelsea, 73, 90, 121, 147, 167
Chiddingford, 13
China, 66, 67, 69, 73, 75, 85, 87
China clay, 90, 93
China stone, 89
Chinese porcelain, 68, 69
Ching-te-chen, 73
Christian VII, King, 141, 142
Churchyard Pottery, 77
Clarke, Alderman, 147
Claw beaker, 8
Clay, 86, 87, 88
Coal-fired furnace, 15

Coalport, 164
Cobalt, 25, 68, 69
Cobalt oxide, 45, 109
Coiling, 103
Colclough, H. J., 150
Cole, John, 158
Cole, Walter, 158
Coloured glass, 24, 44, 45
Coloured pottery, 65–66
Company of Glass Sellers, 16
Cookworthy, William, 74, 75, 167
Copeland, Alderman, 82
Copeland, William, 170
Copeland, William Taylor, 170
Copper, 45
Copper oxide, 45, 109
Copper-plate engraving, 113
Cordalite, 91
Cords, 39, 42
Cornish clay, 86
Cornish (soap)stone, 74, 90
Cornwall, 93
Cousteau, Jacques, 61
Cracking-off, 37–38
Crete, 60, 63, 64
Cristallerie Lalique et Cie, 153
Cristallo, 10, 13, 44
Crizzling, 16, 40
Crown House Ltd, 144
Crown Staffordshire, 165
Crutched Friars, 13, 14
Cullet, 26
Cups, 97
Cut crystal glass, 160–163
Cut glass, 17, 23, 24, 42, 43, 47–57
Czechoslovakia, 24, 44, 45, 54, 139–141

Dali, Salvador, 158
Danneskiold-Samsøe, Henriette, 152
Darling figurine, 145
d'Artigues, M., 138
Dartington, 167
Davies, Peggy, 122
Davis, William, 168
Dawn, 122
Decanters, 19
Decolourizers, 4, 26, 27
Decoration, 5, 10, 12, 17, 43–59, 81, 101, 105–119
  and price, 107–108
  in-glaze, 110

Delftware, 70, 89, 92
Dema Glass, 143–151
de Montmorency-Laval, M., 137
Denmark, see Royal Copenhagen
Dennis Glassworks, 144
Dennis, Richard, 122
d'Entrecolles, Père, 73, 74, 75
Derby, 147
Derby Crown Porcelain Company, 149
Detergents, 134
Devon, 87
Diamond-cutting, 17
Diamond faceting, 163
Donathello, 157
Dorset, 87
Dossie, Robert, 168
Doughty, Dorothy, 122, 169
Doulton, Henry, 145
Doulton, John, 144
Dresden, 71, 156
Duesbury, William, 147
du Paquier, Claude, 71
Dutch East India Company, 70
Dwight, John, 120

Earthenware, 63, 66, 87, 88, 92, 155, 165, 170
Edge-melting, 38
Edinburgh Crystal, 143, 144
Edward VII, King, 143
Egermann, F., 141
Egypt, 1, 2, 3, 5, 63
Egyptian black, 78
Egyptian glass, 45
Eighteenth century, 17–21, 46, 120
Ekman, Johan, 153
Elizabeth I, 13
Enamelling, 23, 43, 44, 107, 128, 164
England, 12–17, 72
English Delftwares, 92, 158
Engraved glass, 160, 161
Engraving, 55–57
Enoch Wedgwood (Tunstall) Ltd, 165
Ensell, George, 21
Erkersreuth Schloss, 157
Etruria, 79
Excise Act, 19, 22
Eyles, Desmond, 122

Faience, 70, 89
Feldspar, 89

Feldspathic stone, 89, 90, 93, 94
Fettle, 100, 129
Figurines, see Pottery figurines
Filigranai, 153
Filter press, 94, 170
Finland, 153
Firing temperature, 88, 100, 103, 110, 114
Fitzpatrick, Bernard, 163
Flashers, 41
Flatwares, 101, 102
Flaxman, John, 79, 80
Flight, Thomas, 168
Flora Danica, 107, 108, 109, 142
Florentine, 164
Fluted pattern, 47, 105–106
Fontainebleau, 4
Foot-boards, 36
Foot-blower, 31, 34, 35
Forest glass, 8–9
France, 72, 73, 90, 93
Franciscan, 165
Frederick the Great, 71
Fry, Roger, 154

Gaffer, 31, 35, 46
Gallé, Emile, 23, 152
Gate, Simon, 153
Gautier, M., 122
Germany, 6, 9, 17, 23, 42, 72, 90
Gibbons, Grinling, 120
Gilding, 43
Glass, history, 1–2
Glass Cathedral, 156
Glass Excise Act of 1745, 18
Glass-blowing,
  birth of, 5–6
  invention, 5
Glass-cutter, 49
Glass-cutting, 47
  mechanized, 51
Glass-making,
  discovery of, 2
  materials, 2–3
  problems, 4–5
Glassware, care of, 135–136
Glaze scratching, 132
Glazing, 63–65, 67, 88, 92–93
Glengary, 46
'Glory hole', 38
Glost kilns, 67

Index

Goddards, 165
Gold addition, 25, 45
Gold band decoration, 102
Gold-enrichments, 114–115
Gold paste decoration, 107, 117–118, 147
  damage to, 134
  prices of, 118
Graal glass, 153
Grafton China Works, 151
Great Exhibition, 22–23
Greece, 63, 82
Greene, John, 16
Griffiths, Eric, 122, 123
Gropius teaset, 157
Gropius, Walter, 158
Grossalmerode, 28
Growan, 75
Guilding, 23

Hald, Edward, 153
Hamilton, Sir William, 6
Hammersley, 170
Hand-burnishing, 115
Hand-made glasses, 42
Hanley, 154
Harcourt, 138
Heat effects,
  on pottery, 133
  on glass, 136
Heath, John, 147
Height defect, 40
Henk, John, 122
Hennezels, 13
Henzey, 159
Herculaneum, 76
Hill, John, 162
Holland, 12, 17
Hollowares, 93–103
Holmegaard, 152
Honey, W. B., 72, 73
Honeybourne, 159
Hop Ware, 159
Houxs, 13
Hughes, James Leigh, 115, 146

Idar Oberstein, 6
Imari, 147, 149
Impurities, 4
Indian Tree, 170

Industrial Revolution, 81
In-glaze decoration, 110
Inspection, 52–53
Inspector's role, 118–119
Intaglio cut glass, 23, 53–55, 160, 161
Irish Glass Bottle Company, 163
Irish cut glass, 18–21
Iron oxide, 109
Iron removal, 94
Irons, 31
Ironstone, 82, 165
Irving, Hugh, 151

Japan, 85
Jasper ware, 76–81, 87, 89, 105, 163
Jenkinson & Co, 144
Jewellery, 79
Jigger, 97
Johnson Brothers, 165
Jones, Alfred B., 151
Joseph, Prince Max, 71
Juliane Marie, Queen, 141

Kändler, Johann Joachim, 121
Kaolin, 67–68, 71, 75, 87
Kaolinite, 86, 88, 89
Kean, Michael, 147
King's Lynn Glass Ltd, 167
Kismet, 155
Knossos, 60
Korea, 85
Kraka glass, 153
Krog, Arnold, 99, 143

Ladre, 120
Lalique, Marc, 153
Lalique, Rene Jules, 152
Lambeth, 146
Latticinio, 12, 23, 43, 153
Leach, Bernard, 85, 87
Lead content, 25, 27
Lead crystal glasswares, 16, 25, 167
Lead glazes, 92
Lead oxide, 16, 27, 28
Lehr, 19, 20, 21, 38–39
Leibnitz, 70
Leith, 144
Levien, Robin, 156
Liberty and Co, 154

Light, Hubert, 146
Lime, 4, 25
Lidner, Doris, 169
Lindstrand, Vicke, 153
Litherland, William, 149
Lithographic printing, 111–112, 164
Lloyd, William, 6
Lech Aline, 4
London, 13, 16
Lopsided wares, 40
Lorraine, 13
Ludwigsburg, 71
Lund, Benjamin, 74, 167

McGrath, Joseph, 163
Machine-made glasswares, 41–42
Madonna, 120
Maiolica, 70, 89, 92, 122, 146
Mansell, Sir Robert, 15
Marvering, 5, 31, 33
Mason, Charles James, 82, 165
Mason, Miles, 82, 165
Mason's Ironstone, 82, 165
Meakin, J. & G., 165
Mechanical aids, 81
Mechanical boy, 31, 34
Mechanical shock, 134
Meissen, 71, 95, 109, 121, 156, 168
Mesopotamia, 2
Middle East, 64, 66, 69
Millefiori decoration, 12, 23
Millefiori paperweights, 12
Mills, Richard, 160
Minton, 84, 85, 96, 101, 102, 107, 115, 117, 118, 122, 146
Minton, Herbert, 146
Minton, Thomas, 146
Mixing process, 27–28
Models, 98, 99, 145
Moil, 38–39, 42
Moore Brothers, 149
Moore, Henry, 158
Moorstone, 75
Morley, Francis, 165
Moser Karlovy Vary, 141
Mould marks, 40
Moulds, 31, 32, 47, 97, 100, 105, 106, 123
Murano, 10, 23

Nailsea glassware, 12
Nailsea jug, 11

Natural glass, 3–4
Needham and Kite, 170
Newcastle-on-Tyne, 13
Newcastle-under-Lyme, 82
Nickel, 26
Ninteenth century, 21–22, 81–85, 121
Nineveh, 64
Nitira, 140
Noke, Charles, 145
Norton Park Glassworks, 144
Nový Bor, 140, 141
Nuutajarvi Glass, 153
Nymphenburg, 71

Obsidian, 3
Old Country Roses, 150
Orrefors, 153–154
Orvela, Heikki, 153
Oven-to-table wares, 135, 146, 155
Oy Wartsila AB, 153

Painting, 47–48, 73, 74, 75, 107, 108, 130, 131
  under-glaze, 109–110
Palissy, 170
Palmqvist, Sven, 153
Paolozzi, Eduardo, 158
Paragon, 151
Parian wares, 82, 170
Paros, 82
Parrish, John, and Co, 160
Pasmore, Victor, 158
Pastes, 18
Pâte-sur-pâte, 76, 84, 85, 117, 146
Pâte tendre, 72
Pearson, S., and Sons, 150
Penrose, George, 162
Penrose, William, 162
Percival, Thomas, 15
Perfection, 138
Perrins, Charles William Dyson, 168
Persia, 68
Petrie, Sir Flinders, 2
Petuntse, 71, 74, 75, 89
Pew-groups, 121
Phaestos, 60
Phillips, Edward, 149
Phoenicia, 5
Pidcock, John, 159
Piercing, 107

Planche, Andrew, 147
Plaster of Paris, 99
Plate-making, 96, 97
Platts Glasshouse, 143–144
Pliny, 2
Peděbrady, 141
Polishing, 51–52, 102
Polychromatic designs, 75
Polychromatic painting, 83
Pompeii, 76
Poole, 88, 154–156
Porcelain, 89–90, 169
 black, 90
 caring for, 132–136
 discovery of, 67–70
 frit, 72
 hard-paste, 70–72, 90, 91, 92, 100
 ingredients, 67
 oven-proof, 135
 soft-paste, 72–76, 90, 121
Porcelaine de France, 72
Porcelaine noir, 90
Portland, Duke of, 6
Portland Vase, 6, 8, 23, 45, 79
Perzellanfabrik F. Thomas, 156
Potash, 4, 12, 25, 26
Potassium carbonate, 59
Pots,
 filling, 30
 firing, 30
 making of, 28–29
 replacing, 29–30
Potter, Beatrix, 150
Potteries, 87
Pottery,
 as art, 61–67
 before the wheel, 63
 caring for, 132–135
 decoration, 64, 68
 definition, 87
 discovery and development of, 60–85
 earliest, 60
 early uses, 61
 essence of process, 61
 preparing the ingredients, 93–95
Pottery figurines, 120–131, 169
 animals, 122–123
 designer role, 123
 history, 120
 materials and methods, 125–130
 moulding techniques, 123
 painting, 131

Poulson, Joseph, 146
Poulson, Samuel, 146
Pownall, William, 146
Price,
 and decoration, 107–108
 justification, 58–59
 of gold-paste decoration, 118
Printing, 110–112
Prisoner of Love, 121
Prozellan, 71
Pugging, 95
Purification, 27

Quartz, 3, 68, 90
Queen's Ware, 77, 78, 106, 163
Queensberry, Lord, 158

Ravenscroft, George, 16, 17
Raw materials, 25–26, 59, 85, 94
Red clays, 87
Red House Glassworks, 160
Red lead, 25, 26, 59
Refractive index, 93
Refractory pots, *see* Pots
Rejection, 39–41
Renaissance, 76
Renaut, Antoine, 138
Ridgway, John, 154
Robinet, 138
Rock crystal, 3–4
Roman Empire, 12
Rome, 6–9
Rosenthal, 103, 107, 115, 156–158
Rosenthal, Philip, 156
Rosenthal, Philip, Jr, 158
Rosenthal Technik, 158
Rotelec Kiln, 149
Rouen, 72
Royal Adderley Floral, 150
Royal Albert, 149
Royal Brierley, 25, 52, 56, 159–160
Royal Copenhagen, 90, 95, 99, 106–109, 120, 122, 124, 125, 128, 141–147
Royal Crown Derby, 147–149
Royal Danish Porcelain Manufactory, 141
Royal Doulton, 94, 99, 120, 122, 125, 127, 129, 130, 131, 144
Royal Grafton, 151

Royal Stuart, 40
Royal Worcester, 73, 74, 84, 90, 122, 167–170
Ruby-red glass, 45
Ruskin, 23, 43
Rye Pottery, 158–159

Saggers, 100
Saint-Cloud, 73
St Mary's Works, 149
Salt glaze, 92, 120
Sampson Hancock, 149
Sand, 4, 6, 10, 25, 27
Sand-blasting, 57–58, 152, 161
Scandinavia, 24, 43
Scratching, 133
Seconds, 52
Seeding, 15, 39
Semi-lead glass, 27
Serviter, 31
Seventeenth century, 120
Sèvres, 73, 84, 117, 121, 168
Shearing, 40
Sherborne, 155
Shone, Pauline, 170
Shrewsbury Green, 117
Shower, 39, 41
Sicily, 6
Sidens, 40, 41
Sidon, 5
Silesia, 141
Silica sand, 25
Silliman, Benjamin, 33
Sillimanite, 33
Silver marks, 135
Silvers, Joseph, 160
Skyphos, 65
Slab pots, 104
Slip, 95, 100
Slipware, 89
Småland, 153
Smith, Constantine, 147
Smith's Blue, 147
Soap effects, 134
Soda, 4, 12, 25
Soda glass, yellowing, 136
Soda lime glass, 25
Solon, Louis Marc, 84, 146
Speckles, 40
Spode, 90, 169–170
Spode, Josiah, 167, 169

Spode, Josiah, Jr, 170
Sponge, 100
Sprigged green hops, 159
Sprue hole, 100
Stabler, Harold, 155
Star China Company, 151
Stemware, 10, 19, 32, 46, 138, 161
Stevens and Williams Ltd, 22, 28, 159–160
Stevens, William, 160
Stoke-on-Trent, 82, 87, 151, 169
Stöltzel, Samuel, 71
Stones, 39
Stonewares, 66, 79, 88–89, 155, 170
Stourbridge, 13, 22, 24
Strass, Josef, 18
Streamline, 155
Stuart, 22, 24, 46, 47, 52
Stuart Crystal, 160–161
Stuart, Frederick, 160
Stuart, Ian, 160
Sunlight effect on glass, 136
Surface imperfections, 40
Susa, 64
Sweden, 153
Syria, 1

Taker-in, 31, 37
Taker-off, 39, 41
Taker-out, 41
Tax-liability, 22
Tea stains, 135
Teapot, 98, 100, 150
Teardrop, 46
Teasers, 30
Tell el Amarna, 2
Terracotta, 88
Thelwell, Boris, 15
Thietrys, 13
Thomas, 42, 156
Thomsen, Christian, 124
Thrower, Frank, 167
Throwing, 103
Tin glaze, 84, 92
Tinning-up, 50
Toby jugs, 122, 146
Tools and techniques, 31–32
Tow, 100
Trailed decoration, 5, 10, 23, 43, 45
Transfer printing, 82, 83, 110–111
Trident Pottery, 62

Triton, 122
Tschirnhaus, 121
Turntable, 41
Tutbury, 151
Tuthmosis III, 1
Twentieth century, 85

Uneven distribution, 40
Ur, 2

Venerable Bede, 12
Venetian glass, 16, 23
Venice, 9–12, 13
Venus, 121
Verzelini, Giacomo, 13
Vienna, 71
Vilhelmsson, Eskil, 167
Vitrea, 141
Vitrearius, Laurence, 13
von Tschirnhaus, Ehrenfried Walther, 70
Vyse, Charles, 145

Waldglas, 9
Walker, James, 154
Wall, John, 168
Water effect on glass, 136
Water removal, 94
Waterford, 18, 162–163
Webb, Charles, 144
Webb, Corbett Ltd, 151
Webb, John, 143
Webb, Thomas, 143
Webb, Thomas Wilkes, 144
Webb, Walter Wilkes, 144
Wedgwood, 87, 105, 106, 163–164, 167
Wedgwood, Josiah, 76–81, 163
Wedgwood II, Josiah, 163
Weight tolerance, 40
Westminster Abbey, 13
Wheel dressing, 49–50
Wheel-engraving, 23
White clay, 87
White glass, 45
White glaze, 68
White House Glassworks, 151, 160
White pottery, 68
Wild, Isaac, 149
Wild, Thomas, 149
Wild, Thomas Clarke, 149
Williams, Samuel Cox, 160
Williams Thomas, Lieutenant-Colonel R. S., 160
Willow Pattern, 82, 83, 111, 146, 150
Wine-glass bowl, 32–35
Wine glasses, 12, 19, 20, 24, 43, 46
Wingen-sur-Moder, 153
Wirkkala, Tapio, 158
Web foot, 40
Wood, Ralph, 122
Wordsley, 143
Würtemberg, Duke of, 71

Young, James Radley, 154

Zeleznéy Brod, 140
Zouch, Sir Edward, 15